I0190011

SOUND & ACOUSTICS HANDBOOK

Written in laymen's language for non-technical people who must use and operate sound equipment

BY
WILLIAM ALDERMAN

Copyright © 2007 by William P. Alderman
Oltermann@hotmail.com

All rights reserved. No part of this book may be reproduced, stored in a retrieval system, or transmitted in any form or by any means except for the inclusion of brief quotations in a review, without the prior written permission by the publisher.

ISBN 978-0-9798376-0-9

Printed in the USA by
Booksurge Publishing
www.booksurge.com

Contents

PART THREE - SOUND SYSTEM MICROPHONES

PART FOUR - SOUND SYSTEM ELECTRONICS

Illustrations

PART ONE - ACOUSTICS

PART TWO - SOUND SYSTEM LOUDSPEAKERS

PART THREE - SOUND SYSTEM MICROPHONES

PART FOUR - SOUND SYSTEM ELECTRONICS

GLOSSARY

APPENDIX 1

APPENDIX 2

APPENDIX 3

FOREWORD

As a sound contractor I designed and installed many sound systems. Following a sound system completion, a training session would be conducted for operator personnel who would be in charge of running the system. Training was a one-day crash course that would familiarize operators with the system design. Operators were assumed to know basic control concepts and techniques, and with this day's instruction be expected to perform competently.

Unfortunately, we were often presented with people who had no previous experience and demonstrated little knowledge. This usually resulted in the operator learning just enough to get by as long as the system was utilized in a routine manner. There were no publications known that would instruct in the basic philosophy and techniques for sound systems that we could give them. All available texts were written for those with a technical background who could understand the technical language and mathematical explanations.

This book with many illustrations fills that void by discussing the basics of acoustics and sound in easy understood language. A prerequisite to understanding a sound system's operation is an understanding of the acoustic environment in which the system is being used. That is why this writing starts with this subject.

After acoustics is discussed, loudspeaker sound distribution is considered. Next microphone selection and sound pickup techniques are examined. Lastly, mixer controls are explained with suggested adjustment approaches. Also graphic equalizers, compressors, and limiters uses are defined with setup instructions.

Sound system operators (often called technicians and even sometimes called engineers) come and go leaving the sound system operation to someone who needs training. Churches with volunteer help suffer the most. The availability of this book to the novice will be most helpful and to the competent a handy reference.

INTRODUCTION

A photograph is only a two-dimensional paper and ink representation of an area of three-dimensional objects. Light, shadow, and color affect the quality of this visual illusion. As amateur photographers, we analyze our environment of light and color only to find that our techniques, at best, yield mediocre results. All of us with normal sight see as well as the artist or the professional photographer, but the reason we cannot faithfully create that illusion on paper or canvas is because we do not understand what we see. The expert must know how to properly use and manipulate light, color, and shadow to achieve a good outcome.

As with photography, most of those who use sound equipment to amplify, record, or in some way enhance sound, do not understand the nature of sound in given acoustical environments. Having the finest sound equipment will not always give adequate results if the operator is unacquainted with acoustical principles. A professional photographer understanding visual surroundings can take fairly good pictures even when using a cheap camera.

A good photographer uses his or her camera to make pictures that draw your attention to the subject matter while eliminating any technical flaws that will distract you. In the same manner, a good sound system is best when it draws your attention to the performance without distracting you by its utilization.

Too often pastors, music ministers or church sound system operators are responsible for purchasing sound equipment for their various audio applications. Frequently they rely solely on the advice of vendors or salespersons as to what they should buy. While sound equipment suppliers may well understand the electronic aspects of their equipment, many know little about acoustical conditions. As a result, equipment is purchased and installed that does not work well in its environment. Inadequate performance of a sound system can result from one of four things, or the combination of them:

- The sound equipment in the system may not be comprised of the right components.
- The sound equipment may not be properly installed.
- The sound system operator may not have sufficient audio technical skills.
- The performer may not use proper or professional techniques.

Unfortunately, modern man must often use sound equipment to speak or broadcast to large groups of people. In this book, I will show in simple, easily understood, laymen's terms, how to comprehend your acoustical situations and properly utilize sound equipment to your advantage. This is written with both the sound system user and sound system operator in mind.

Part One
ACOUSTICS

THE NATURE OF SOUND

One of the keys to comprehending sound behavior is the understanding of tones. The sounds we hear naturally about us are always complex tones. A complex tone is a pure tone that determines the pitch, blended with other tones to give this tone a unique texture and color called timbre. That is why the same note when played on different musical instruments produces a different sound, although the pitch is the same. Consider the fact that a complex high pitched tone also contains low tones and a complex low-pitched tone also contains high tones. This enables the tone controls on your radio to make high sounds "bassy" by accenting the inherent low tones and low sounds "brassy" by accenting the inherent high tones. Rooms can also affect the tonal quality of sound. Spaces heavily carpeted and draped can cause sound to be bassy and heavy. That same sound in a room with exposed hard walls, floor, and ceiling would sound brassy and thin.

Nowhere in nature can you hear a single pure tone. All the tones come to us with a blend of other tones. Not until this modern era were pure tones artificially produced using electronic devices. The tone that accompanies the TV channel test pattern before and after programming is a pure tone. Pure tones lack the texture and color that complex tones have, and listening to them is not very pleasing. Tones are the sensation produced through the ear when certain vibrations are set up in the surrounding air by a vibrating (sound) source. Sound may also be considered to be the vibrations themselves or the vibration energy that produces these vibrations. Sound has two important characteristics: *frequency* which is the number of vibrations per second representing pitch, and *amplitude* or intensity which represents loudness. Once tones are emitted into our environment, they can do several things:

- Go in a straight line until they dissipate in air.
- Reflect off surrounding rigid surfaces.
- Echo off perpendicular hard surfaces.
- Be absorbed by surrounding objects.
- Resonate in a room in the same way sound resonates within the sound box of an acoustical musical instrument.

THE NATURE OF SOUND OUTSIDE

Let us start by discussing how sound behaves outside in the simplest acoustical setting. If you were standing in a quite open

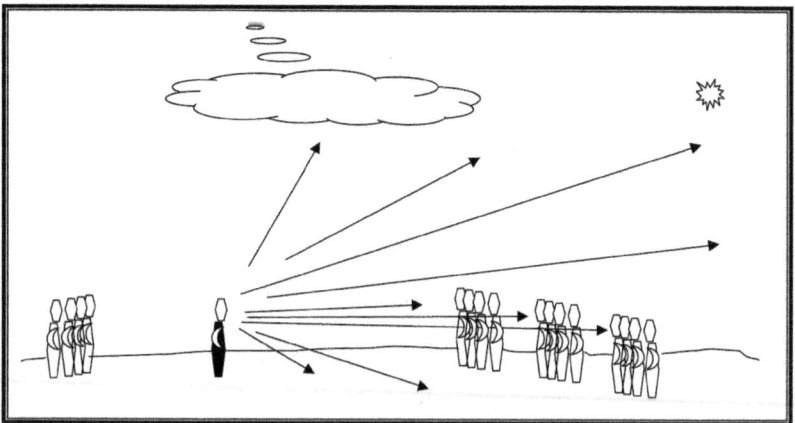

OUTDOOR ABSORBENT ENVIRONMENT - Only those in the direct sound path will hear the person speaking. There is nothing to reflect the sound back to them standing behind. All the sound projected will absorb into the earth or dissipate into the air.

grassy field without any nearby buildings or hard surface structures, and without the wind blowing, you would be in an environment where almost all sound is absorbed. An absorbent acoustical environment is where sound has no place to reflect or diffract (bend). Here you can only hear sound that comes directly from the source. If someone should speak to you with their back

SOUND PROJECTION ON A COLD DAY - Sound projected over a cold surface on a cold sunny day will attempt to rise but warm air above will bend the sound back down.

to you, you would not hear them. In this natural setting, sound energy is absorbed by soil, grass, foliage, and air.

Items that will reflect, bend, or bounce sound in a natural setting are tree trunks, rocks and water. Temperature and wind can affect sound movement too. On cold sunny days when the ground is cooler than air, sound will tend to skip across the ground. As the sound rises above the ground, warmer air will bend it back down. Conversely, on warm days rising air currents will bend sound upward. Wind can also act on sound. Sound in the direction of wind can travel longer distances than normal. Sound projected in to the wind will be diffracted (bent) upward causing normally heard sounds at a distance to not be heard.

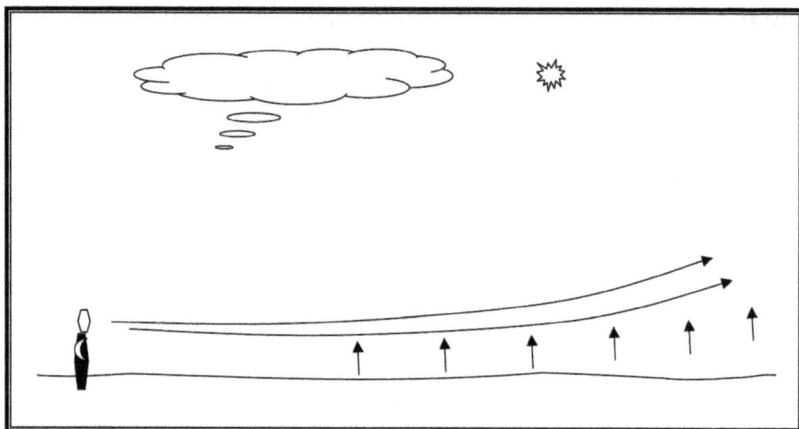

SOUND PROJECTION ON A WARM DAY - Sound projected over a distance will tend to bend upward because of rising air currents.

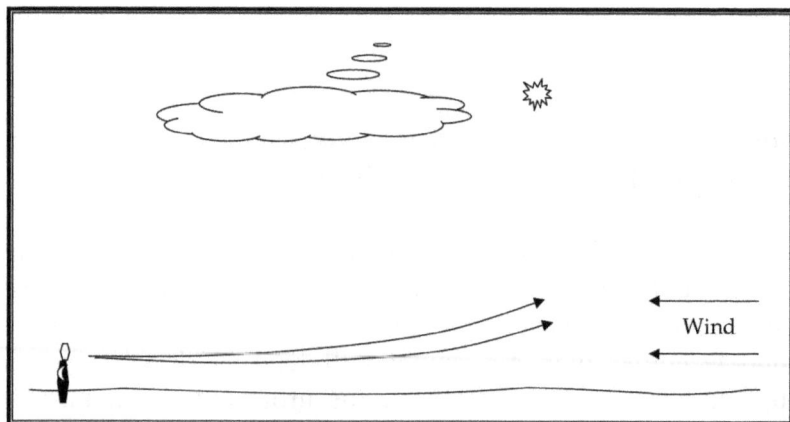

SOUND PROJECTED INTO THE WIND - Outside, sound projected against the wind will bend upward.

The mid-range or mid-tones of sound produced in an open grassy field will diminish to ½ the loudness each time the distance is doubled from the source. The loudness at 10 feet will diminish to ½ the loudness 20 feet beyond the 10 feet, that is, at 30 feet. In the same manner, loudness at 30 feet will diminish ½ the loudness 60 feet beyond the 30 feet, that is, at 90 feet. In the days before sound systems with microphones and loudspeakers, persons speaking would stand on hard pavement, before walls, or even a stone cliff in order to be heard better. These sound reflective structures served as sounding boards.

Stage shells or sounding boards can double the loudness of sound to the open field. Spoken sound moving away from the audience is redirected by reflection to the listeners. This sound combined with the direct sound amplifies the sound so that the person speaking can be heard more loudly.

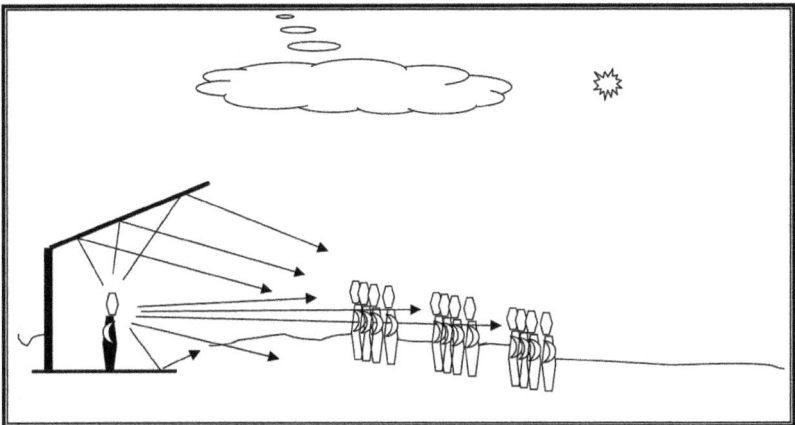

OUTSIDE STAGE SHELL PROJECTION - When a person speaks in an outside stage shell, the shell will reflect the upward moving sound back to the audience to amplify the sound.

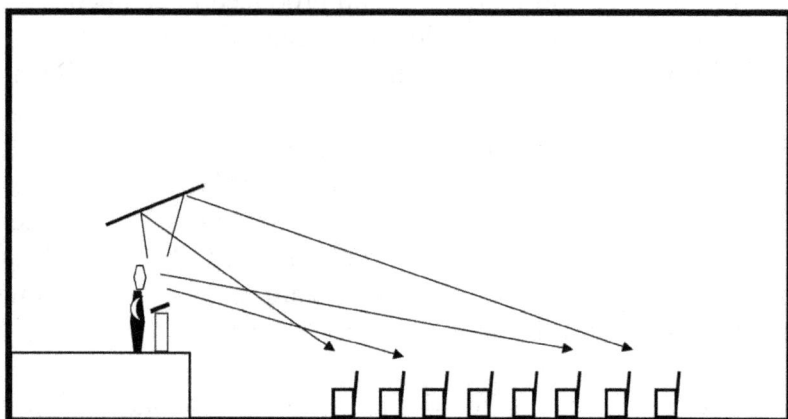

SOUNDING BOARDS - Before sound systems, sounding boards were use in large buildings to reflect sound to the audience.

Often, sounding boards were used in auditoriums and church sanctuaries to help project the sound better. The Bible records that Jesus addressed large numbers of people away from the

SOUND PROJECTED FROM BOAT OFF SHORE - Here is an example of how Jesus spoke to crowds at the seaside. He moved the boat out into the water and took advantage of sound reflecting off the still water and rising air currents to carry the sound up the hillside.

noise of towns. To speak to the crowds, He would often step into a boat suggesting that He was at the bottom of a hill. Then He would move the boat out into the water and take advantage of rising air currents and sound reflecting off the still water to carry the sound up the hillside.

Sounds Traveling Over Distances

When attending a holiday parade, a brass band in the distance will sound as if it is comprised of bass drums. As it approaches, the higher-toned instruments start to be heard. The closer the band gets, the brighter the sound.

There are several factors that cause this phenomenon. Most high-toned instruments in a brass band by their very design project directionally while bass instruments do not. Projected low tones diffract (bend) more easily and travel more places being less obstructed. While all tones travel the same speed, they do not all travel uniformly. High tone intensity will diminish in strength faster while traveling through the air than low tones.

Things such as buildings, trees, the street, temperature gradients, humidity and wind are all factors that act upon sound transmission.

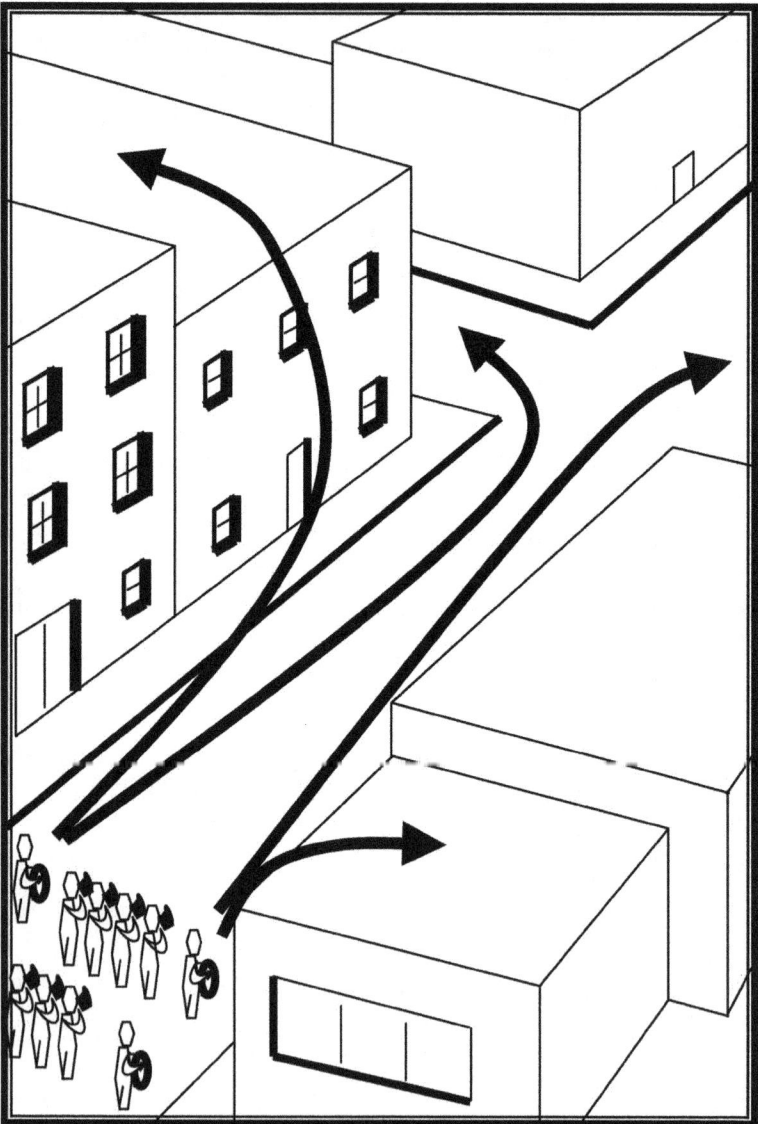

OUTSIDE BASS INSTRUMENT PROJECTION - This illustrates how sound from a bass drum bends (diffracts) around corners. This, coupled with the fact that the strength of high tones dwindle faster, is why the bass drum and bass instruments are heard first from an approaching band.

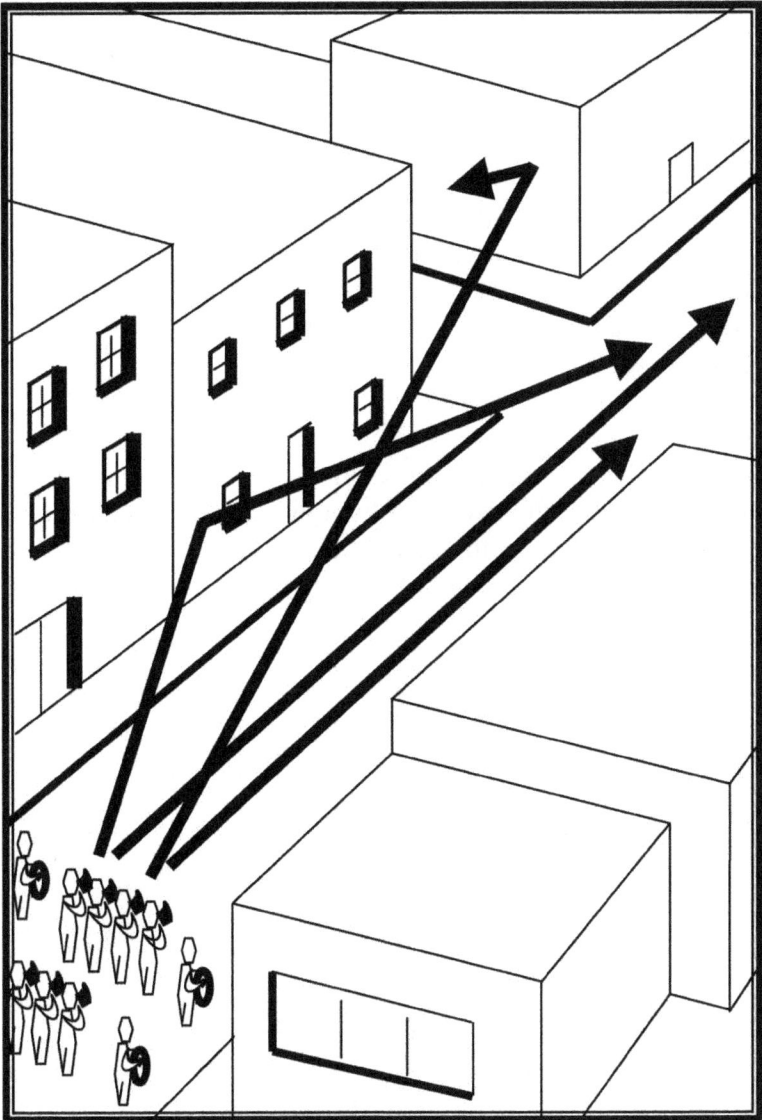

OUTSIDE TREBLE INSTRUMENT PROJECTION - This illustrates how sound from instruments above the bass range project. Notice that the sound goes straight and reflects off hard surfaces at the same angle it strikes.

THE NATURE OF SOUND INSIDE

We have discussed sound outdoors starting with a grassy field environment where the sound is almost totally absorbed. From this extreme, consider the other extreme where an enclosed room made of rigid surfaces is almost totally reflective. Almost none of the sound energy in this room will be immediately absorbed. When sound is produced in this room, it will travel much by bouncing from hard surface to hard surface until it dissipates in the air. This is called sound reverberation and reverberation is the opposite of absorption.

In the following illustration we look down on the room to see how sound will reflect in a horizontal plane. Imagine this same room viewed from the side. You would also see sound reflecting in the vertical plane in the same way. If this room were full of people, few would understand the words spoken.

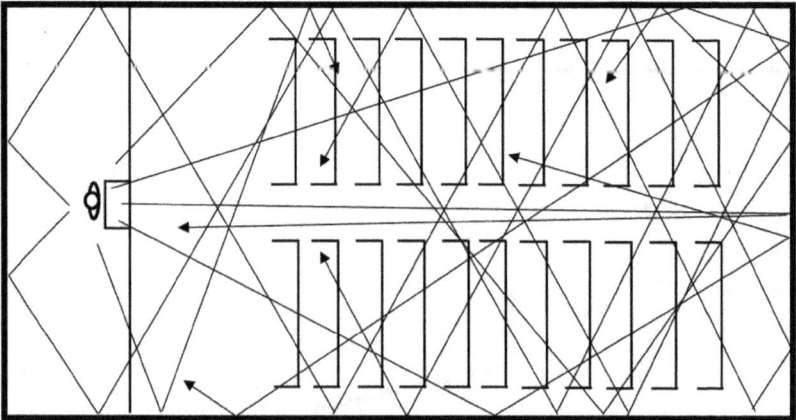

ROOM WITH HARD WALLS - Bird's eye view of a person speaking in a room with hard walls. This illustrates how sound will reverberate by multiple reflections.

In order to hear sound distinctly in the highly reverberating acoustical environment you must be close to the person speaking. If you cannot hear their voice directly, you will hear, but the sound will be unintelligible. Likewise in order to hear sound in the highly absorbent acoustical environment you must be close to the person speaking. If you cannot hear their voice directly, you cannot hear them at all. As we can see, neither a totally absorbent nor a totally reflective location provides a proper setting for projected sound to be heard. When thinking about what comprises a good acoustical sound environment, it is good to go back to see what worked outdoors. Remember that the person speaking stood in an area where sound was reflected by the stage shell and the listeners sat in an area where the sound was absorbed. If we start with a room that is designed to simulate that outdoor condition, then we have a room that will approach the acoustical conditions desired. Notice in the following illustration that one half the room is reflective like an outdoor stage shell and the other end absorbent like the outdoor grassy field. Of course, a real room is never all absorbent in one half end and all reflective in the other.

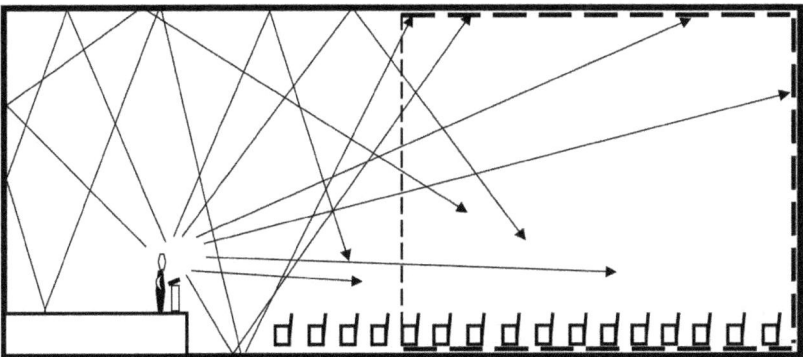

ROOM HALF REFLECTIVE AND HALF ABSORBANT - This room is designed to simulate the stage shell in an open grassy field. The performer's half is sound reflective and the listener's half is sound absorbent.

Also, when fixtures, furniture, windows, doors are added to the room, the parameters change. The outdoor model of a stage shell in an open grassy field is a great way to show highly reflective and highly absorbent environments together being used to an advantage. With this in mind, we have the mental framework to help us understand the basics for proper sound transmission in a room.

When we look at the acoustical properties of a room, there are seven factors that must be considered. Sound in a room will usually encounter elements that will reflect it, diffract (bend) it, echo it, reverberate it, resonate it, and absorb it. Also, rooms of certain dimensions will set up standing waves. Please note that I use the word "room" in a general sense. Room can refer to an auditorium, a church sanctuary, a meeting room or any other similar enclosure.

Early Reflection
A room with sound reflections in the area of the performer such as found in an outside stage shell can double the loudness of projected sound. These are referred to as "early reflections" because they bounce one to a few times before reaching the listener. These early reflections arrive to the listener so rapidly that they combine with the direct sound and are not heard distinctly.

Diffraction
This is the bending of sound. It is not often a concern in the average room where the sound is contained. Remember that low tones bend much more easily than higher tones.

Echo
Echo occurs when sound bounces directly back from a distant perpendicular hard surface creating a 1/20 of a second or longer sound behind the direct sound. This reflection of sound is often

referred to as "slap back". Seldom is echoing a good thing because it confuses the direct sound, especially the spoken word.

Reverberation

Reverberation is sound that is reflected many times off multiple surfaces such as hard walls, a hard floor, a hard ceiling, and hard objects within the room. Reverberating sound is the diminishing sound that remains after the direct sound is heard. The size of the room and the absorption factors of the walls, ceiling, floor, and furnishings govern the amount of reverberation time. The length of time of the reverberation in a room is important to know. Expensive instrumentation can be used or you can create a sharp sound when the room is quite and listen to it. I have seen people shoot blank pistols, slam large books shut, pop paper bags, and clap their hands sharply to create this sound. By counting or using a stopwatch you can fairly accurately estimate the reverberation time in seconds. One to one and a half seconds reverberation is ideal for most auditoriums and church sanctuaries. A little reverberation is good and gives liveliness to the direct sound. Less reverberation favors the spoken word while more favors the enhancement of music.

Resonance

Rooms often have areas within them that resonate to certain pitches of tones that will create an effect called "hangover". These persistent resonating tones create peaks and valleys in the room characteristics. This tonal discrimination will result in places in the room sounding hollow.

Standing Waves

When a single sustained tone is projected into a room having parallel surfaces, standing waves may set up. This stream of vibrating air made by the tone projected into the room is called a wave train. The wave train striking a wall will produce a reflected wave train. Standing waves are created when two wave trains moving in opposite directions interfere with each other. Walking through the room produces sensations of an increase and decrease in the intensity of sound. The original and the reflected wave trains combine or subtract their intensities throughout the room resulting in standing waves. While standing waves may be demonstrated with a sustained tone, this is only an indicator of problems that will occur under normal use. Intensity of sound heard in different parts of the room will vary substantially. Constructing a room with non-parallel walls and multilevel ceiling sections can prevent standing waves. There are recommended dimension ratios for rectangular rooms to minimize standing waves. Make the room width 1.25 times wider than the height while making the length 1.6 or 2.5 times longer than the height.

Absorption

There are no conventional building absorption materials made that will absorb sound in its full spectrum of tones like the large quiet grassy field. Nor do building absorption materials absorb all tones at the same rate. A certain material may absorb high tones more thoroughly than low tones or the middle tones more completely than the high tones and low tones. Nevertheless, there is a vast array of materials available in various colors, shapes and designs.

When a professional talks about the characteristics of sound absorbing materials, he will speak of sound frequencies measured in "Hertz" rather than pitches of tones. These two terms refer to the same thing. Pitch "A" above "Middle C" is the sound frequency of 440 Hertz (Hz). Hertz or "Hz" denotes vibrations per second called cycles per second.

Musical Note	Frequency (Hz)	
C8	4186	HIGHEST NOTE ON PIANO
B7	3951.1	
A7	3520	
G7	3136	
F7	2793.8	
E7	2637	THE NOTES OF A MUSICAL KEYBOARD AND THEIR FREQUENCY
D7	2349.3	
C7	2093	
B6	1975.5	
A6	1760	
G6	1568	
F6	1396.9	
E6	1318.5	
D6	1147.7	
C6	1046.5	SOPRANO VOICE Upper Limits
B5	987.77	
A5	880	
G5	783.99	
F5	698.46	ALTO VOICE Upper Limits
E5	659.26	
D5	587.33	
C5	523.25	
B4	493.88	TENOR VOICE Upper Limits
A4	440	
G4	392	BARITONE VOICE Upper Limits
F4	349.23	
E4	329.63	
D4	293.66	BASS VOICE Upper Limits
Middle C	261.63	SOPRANO VOICE Lower Limits
B3	246.94	
A3	220	
G3	196	ALTO VOICE Lower Limits
F3	174.61	
E3	164.81	
D3	146.83	TENOR VOICE Lower Limits
C3	130.81	
B2	123.47	
A2	110	BARITONE VOICE Lower Limits
G2	97.999	
F2	87.307	
E2	82.407	BASS VOICE Lower Limits
D2	73.461	
C2	65.406	
B1	61.735	
A1	55	
G1	48.999	
F1	43.654	
E1	41.203	
D1	36.708	
C1	32.703	
B0	30.868	
A0	27.5	LOWEST NOTE ON PIANO

PROBLEMS IN THE ROOM

Sound arrives to the listener in a room in three different ways. First there is the direct sound, then the early reflected sound, and after that, the reverberating sound. Direct sound is the sound that goes directly to the listener from the performer. Early reflected sound is sound that comes rapidly after the direct sound in the same manner that sound is reflected out of a stage shell outdoors. It is so fast that its delay is not discernible. Lastly, reverberating sound is the lingering diminishing sound that is heard after the direct sound and the early-reflected sound has stopped.

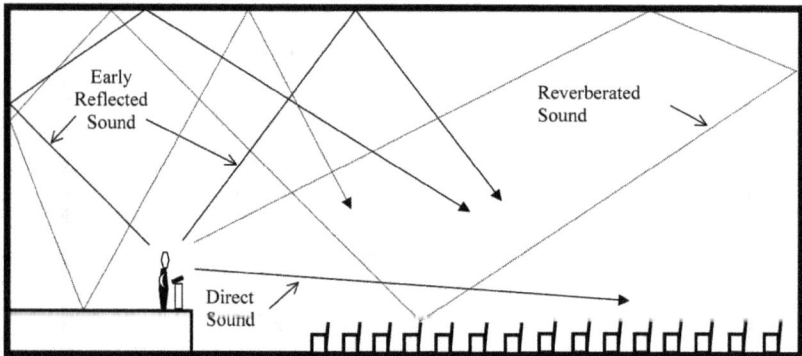

SOUND TRAVEL WITHIN A ROOM - This illustration depicts the direct sound, the early reflected sound and the reverberated sound.

In the following we will discuss acoustical problems that are often incurred. This writing is not intended to make you competent to correct the acoustical structure of rooms, although you may be able to in some cases. But, this is written to help you understand, cope, and adjust with situations at hand and to be able to know how to discuss intelligently the acoustical problems with a professional should that be necessary. Although you know what the problem is, the solution to correction may be very complex, particularly in larger rooms.

Reverberation Problems

When analyzing the reverberation problems in a room, always keep in mind the outdoor model of the acoustic shell in a grassy field to remind you of the important reflective areas and the important absorbent areas in the room. To understand the problem, there are three circumstances that need to be addressed.

1 - Is the tonal quality bassy, normal or brassy?
2 - What is the duration of the reverberation?
3 - What are the structural aspects that cause the reverberation?

First determine the length by creating a loud pop in the performance area and timing the sustained reverberation. Listen to the tonal characteristics of the reverberation. Is the sound boomy? Is it in the midrange of tones? Is it bright and brassy and in the high range? If you have trouble making that judgment, try listening to some familiar recorded music in the room.

If your room is boomy, you likely have a room where the problem is not reverberation, but a room environment that absorbs the mid and upper ranges of sound leaving the low tones. Another possible factor is that the room construction may not be very rigid and the structure may vibrate to the low sound causing the bass to be enhanced. As bass instruments are played inside these kind of rooms, the bass sound can be heard outside the structure. When sound vibrates a structure, a window or any other barrier causing the sound to be emitted to the other side, an effect called refraction occurs. Remember the power in bass tones is much greater than higher tones and that bass tones can more easily bend and seep through cracks and openings.

So, in boomy rooms look for thick carpet and dense fibrous materials that absorb mid and high tones, and for vibrating room structures. Remember that if absorbent materials are to be removed, start in the performing area where hard surfaces are an advantage. Should you determine that some bass control is necessary, you will find merely placing some absorbent covering on the wall will not solve the problem. Bass tones are trapped and absorbed. Usually absorption cavities or tubes are erected in corners and spaced along walls for this purpose. I would suggest here that you contact a professional.

When the excessive reverberation is natural sounding (not bassy or boomy), then thought needs to be given to approaches that do not change the sound qualities of the room. Here are things that need to be considered.

1 – Put absorbent materials in the listening area. You are not likely to need any in the performance area except for side to side ringing that might occur during a production. This happens when sound is produced between two parallel hard walls. Treating one of the two walls at the point where the sound bounces back generally eliminates the ringing.
2 – Use the right materials. An internet search will reveal a large quantity of acoustic sound absorbent materials available. Choose that material that absorbs the tone (frequency) range you need.
3 – Placing absorbent material in the right spots is extremely important and this is where the professional is needed. Having the right amount in the right spots will give you the correct reverberation time. But if you have the time and money to experiment, the back wall in the listening area is where to start. Often two problems can be cured. You may be able to eliminate any back wall echo while lowering the reverberation time. Start at the top of the wall completely covering as you come down. As you continue down the wall, check the reverberation time and stop when the right time is

reached. You may need to cover the entire wall with the exception of where people come in contact with it. Sound absorption materials tend to be fragile and damage easily. After that, if you need more sound absorption, look at the front of the balcony, large pillars, the side walls, the ceiling above the listeners, etc. People in rooms also absorb sound. An empty room will have a longer sound reverberating time than when it is filled with people. Often padded permanent seating is put into rooms to compensate for empty seats. So if padded seating is used, whether the room is full of people or not, the reverberation time will be less affected.

If the excessive reverberation is brassy sounding, approach the problem in the same manner as described for the natural sounding reverberation. Select absorbent materials for that range of tones. While very brassy or bright rooms can be annoying, rooms that tend to be bright are pleasing to most ears. Also, words spoken in a bright room are more easily understood.

Echo Problems
Echo, frequently referred to as "slap back", occurs when sound bounces directly back from a distant perpendicular hard surface. This is often simply corrected by completely covering the opposing wall or structures with absorbent material. However if your room's reverberation time is correct, adding this material may shorten it too much. If this is the case, the echo needs to be diffused rather than absorbed. When you diffuse the echo, you scatter it. This is done by making the surface of the hard echoing wall irregular. Irregular and hard so that the reverberation time of the room is little affected. When creating the irregular wall, don't make the mistake shown in the illustrations. Here at first look, the illustrated wall design

patterns seem right for diffusing and scattering sound, but they aren't. Commercially available diffusion panels are expensive but effective and should be considered when needed.

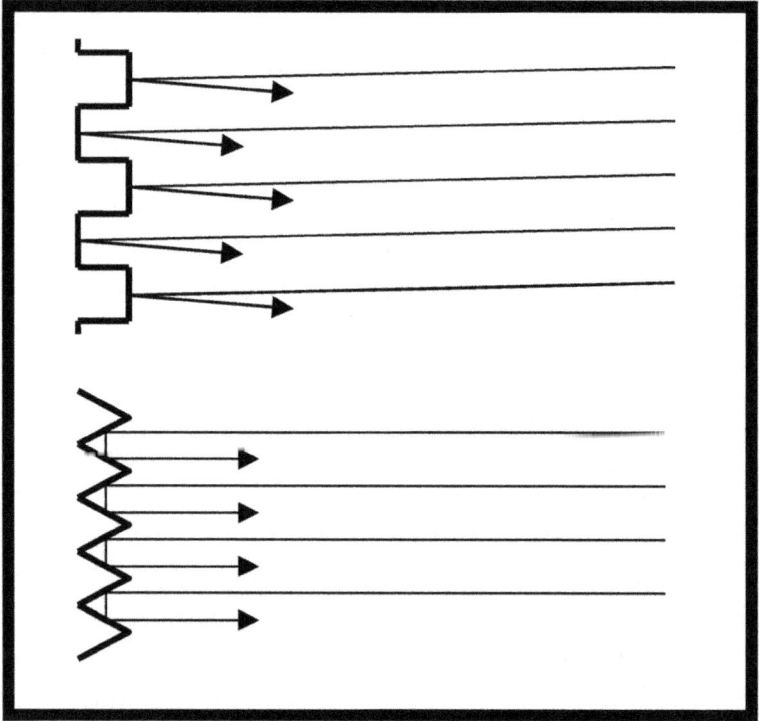

IMPROPERLY SHAPED WALLS - This illustration shows two designs for irregular shaped walls that **do not** diffuse the echo. Note here that the sound is directed right back to the source.

Resonance Problems

Some rooms have areas within them that resonate to certain pitches of tones that will create an effect called "hangover". These persistent resonating tones create peaks and valleys in the room characteristics. This tonal discrimination will result in hollow sounding areas in a room. Room resonating areas are usually alcoves, corners, ceiling domes, and other such cavities. Introducing sound absorption in the area will stop the hangover effect. Often, full treatment is not necessary. Covering one parallel surface will sometimes stop the problem.

Standing Waves Problems

If you wish to test your room, place a loudspeaker on a table where a performer would stand, then produce a sustained single pitched tone in the midrange to be heard all over the room. If standing waves do set up, you will sense an increase and decrease in the intensity of sound when walking through the room. A room that will produce standing waves will have soft and loud spots throughout the listening area under normal conditions.

The approach to correcting this is similar to the echo problem. Standing waves set up between parallel walls. They may set up between the wall behind the performer and opposing back wall or they may set up between parallel side walls. Putting absorption on one of the parallel walls where standing waves set up is usually the solution.

Measuring Sound Distribution in a Room

When professionals desire to evaluate the sound distribution of a room, they produce a sound in the performance area of the room and measure the loudness at many points in the listening area. The loudness or "sound level pressure" is measured by a hand held instrument know as a Sound Level Meter. Sound pressure levels are measured on the dB-SPL scale. The term dB (decibel) is a unit of measure and SPL denotes that Sound Pressure Level is being measured. Outdoors, when the distance from the sound source doubles, the area that the sound expands into quadruples. Since sound energy is distributed over a wider area the sound energy drops to ¼ the intensity, not ½. To our non-linear ears the perception drop is around ½ when there is no reflected or reverberant sound present. Fortunately, the loudness of sound is not directly proportional to its intensity (power). Our ears reduce in sensitivity as the intensity increases and increase in sensitivity as the intensity decreases. A loud explosion may be one trillion times louder than a mosquito buzz, but we don't hear it that much louder. So the term dB is a logarithmic measure rather than a linear measure that better mathematically represents the way we hear. The dB unit is used to measure the relative intensity of sound. A sound 10 times more powerful than another sound is said to be 10 dBs more intense. At 100 times more powerful, 20 dBs more intense. At 1,000 times more powerful, 30 dBs and a 10,000 times more powerful, 40 dBs. The dB measure gives an approximate connection between the intensity of sound and the loudness it causes. A change in intensity of 3 dBs causes the smallest change in loudness that the average human can sense.

The dB-SPL levels are best understood when referenced to the "Typical Sound Levels Encountered Daily" chart (next page). Since the association between sound intensity and how loud we hear is complex, the chart helps us to generally grasp their relationship. On the chart you will see that 0 dB-SPM is not actually zero, but a point of reference equal to the sound pressure at the threshold of hearing. The chart will illustrate the approximate dB-SPL measure in relation to the loudness we hear. Also, our ear does not have the same response to sound of different tones at the same loudness intensity. At low sound pressure levels we hear the middle sound range tones (frequencies) better or louder than the low range and high range.

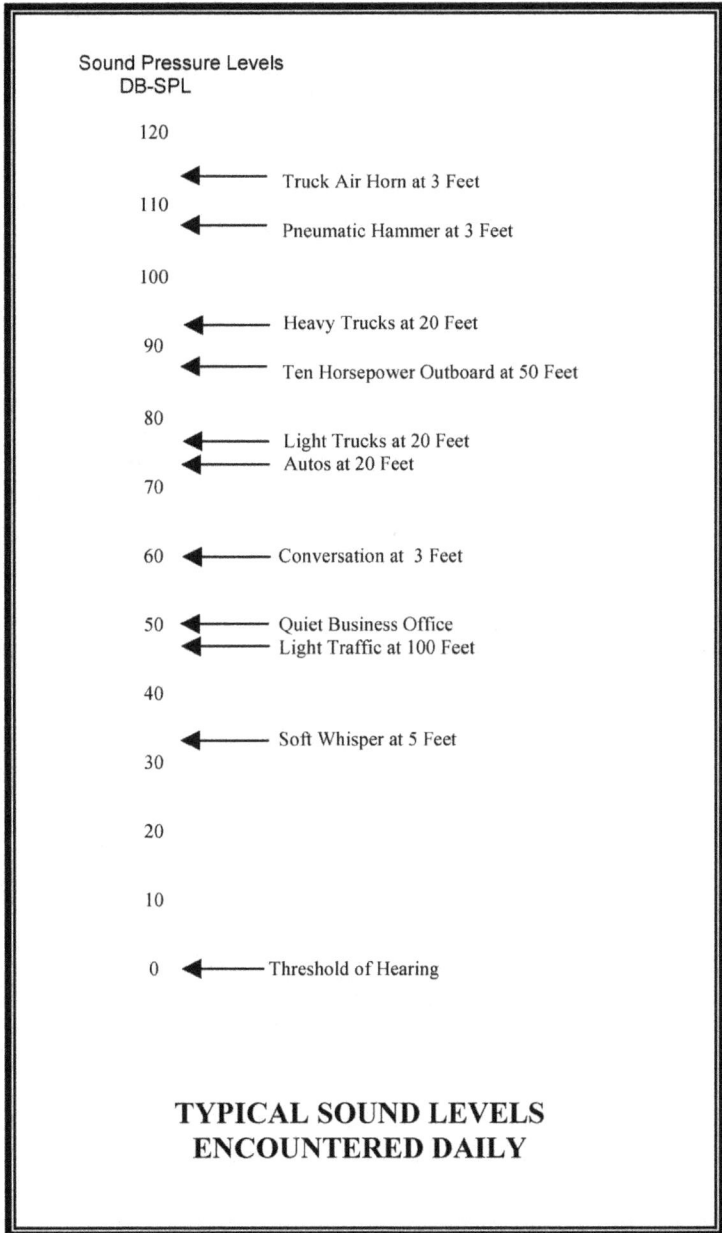

Sound Pressure Levels
DB-SPL

120

110 ◄——— Truck Air Horn at 3 Feet

◄——— Pneumatic Hammer at 3 Feet

100

90 ◄——— Heavy Trucks at 20 Feet

◄——— Ten Horsepower Outboard at 50 Feet

80

◄——— Light Trucks at 20 Feet
70 ◄——— Autos at 20 Feet

60 ◄——— Conversation at 3 Feet

50 ◄——— Quiet Business Office
◄——— Light Traffic at 100 Feet

40

30 ◄——— Soft Whisper at 5 Feet

20

10

0 ◄——— Threshold of Hearing

**TYPICAL SOUND LEVELS
ENCOUNTERED DAILY**

The middle range spans from about C3 (130Hz) to C7 (2,093Hz). As sound volume increases we start hearing the high and low ends louder in relationship to the middle range. At a very high volume you can hear the high and low ranges at nearly the same loudness as the middle ranges. That is why music played at a low volume on the radio sounds better when the sound volume is turned up.

The Sound Level Meter has adjustments called weighted curves to make it measure the way our ears hear sound. The "A" weighted curve adjustment is generally used when measuring sound levels where public performances occur. When a professional makes sound pressure level measurements in a room, a "Pink Noise" is electronically generated through a loudspeaker in the performance area of the room. Pink noise sounds like a waterfall and contains all the tones (frequencies) uniformly produced that are perceptible to man.

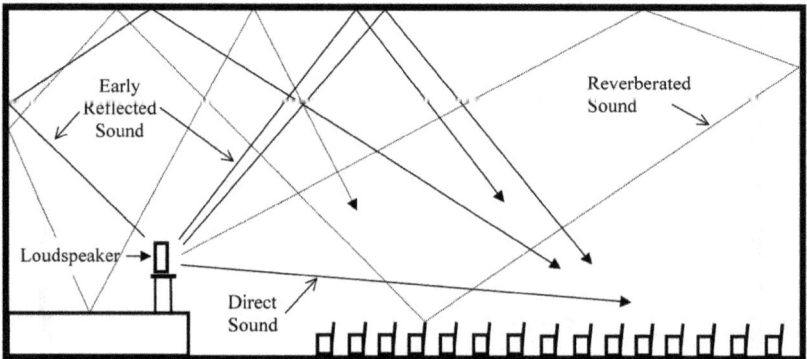

MEASURING SOUND LOUDNESS IN A ROOM - A loudspeaker in the performance area sends out "pink noise" for measuring sound loudness throughout the room.

Sound levels being measured throughout a room will be more nearly uniform in loudness, if the room has no echo, if the room has no standing waves, if the room reverberation is in the range of 1 to 1.5 seconds and if no objects shadow the transmission of the sound. Note, as the direct sound diminishes in loudness, the early reflected sound directed back into the listening area adds to the direct sound to keep the intensity of the loudness for a greater distance.

Seeking Solutions

Even though you know the acoustical problems that exist in a room, the solution to these problems may be very complex. Large books have been written on this subject and acoustical engineering evolves a great deal of higher mathematics. I have discussed here in layman's terms, basic elements to give you a better understanding on how to confer with professionals who can help you with your acoustic needs or your sound equipment acquisition needs. Also, knowing these basic elements will help you to administer better application and operation techniques with your supplied sound system and given environment.

Sound and Acoustics Handbook

Part Two
SOUND SYSTEM LOUDSPEAKERS

As we all know, sound systems detect sound, amplify and process the detected sound, and then distribute that sound to selected listeners. These processes represent the three basic hardware segments in which all sound systems are divided. When the sound system includes loudspeakers that project sound to listeners, we must not forget that the acoustical environment is an integral part of the system. Since this is so, we should always think of the sound system with loudspeakers as an electronic/acoustical (electro/acoustical) system. Electro/acoustical systems that are use to amplify sound for a live performance are called *sound reinforcement systems*.

REINFORCEMENT LOUDSPEAKERS

A sound system designer's most difficult task is choosing the loudspeaker or loudspeakers needed for a reinforcement system. Too often loudspeakers are chosen without proper thought to how they will perform in the environment or where to correctly place them. Selecting the right loudspeaker involves careful consideration of the room size, shape, and listening area. A designer will first attempt to apply the single location loudspeaker approach because it provides greater naturalness than loudspeakers in multiple locations. To understand this approach, consider the outdoor quite open grassy field we talked about in the acoustics section. Remember that the person speaking could only be heard by those in the direct line of sound because of the absorbent environment. However, when a stage

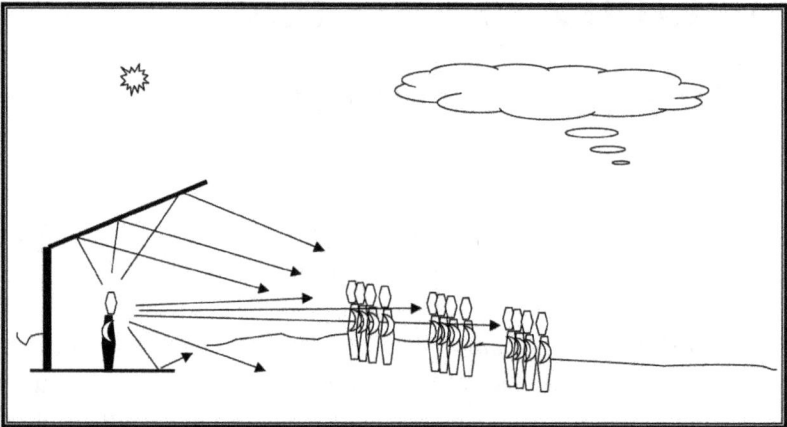

SOUND PROJECTED FROM A SHELL - When a person speaks in an outside stage shell, the shell will reflect the upward moving sound back to the audience to amplify the sound.

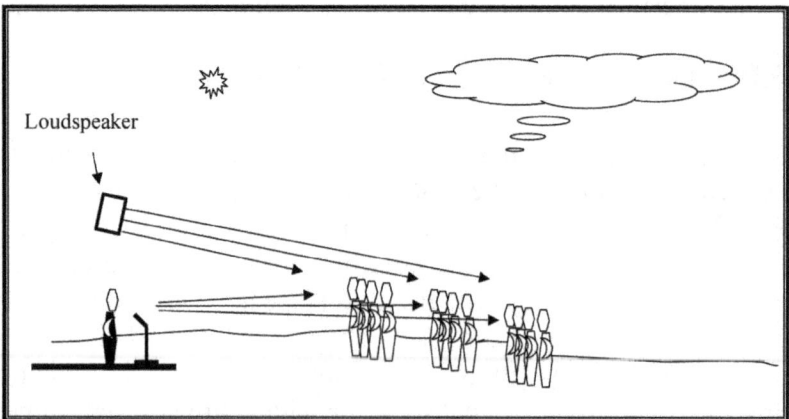

LOUDSPEAKER REPLACING SHELL - The sound system loudspeaker replaces the shell by projecting sound in the same manner back to the audience.

shell was added, the person speaking was heard better and at a greater distance. Now if we substitute that stage shell with a sound system, the sound from the loudspeaker will replace the reflected sound from the shell. A loudspeaker or loudspeakers located above the performer is consistent with the natural flow of sound and provides a natural sound distribution.

One reason that it provides the best distribution is because of the way we hear. Our ears are in the horizontal plane and when a noise occurs around us, we sense the direction from which it comes. But when the sound is from above, the direction is not nearly as discernible. The vertical placement of the loudspeaker above takes advantage of our inability to localize sound sources vertically. We look to the performer using that sound system and sound will appear to come from him although the system sound is coming from above. Also, when the above loudspeaker is located at the center of the room, the reflections from the projected sound will be symmetrical providing smoother distribution and greater loudness.

When trying to determine the proper loudspeaker configuration for a room, several questions need to be ask.

1 - Will the loudspeaker array reinforce speech only or both speech and music?
2 - Where will I locate the loudspeaker or loudspeakers?
3 - What are the coverage angles from the proposed loudspeaker location in both horizontal and vertical planes necessary for adequate sound distribution?
4 - What power capabilities will the loudspeaker array need?
5 - How much electronic amplification power will be necessary?
6 - What kind of loudspeaker or loudspeakers will comprise the configuration?

Cell And Sectoral Type Loudspeaker Horns

Before 1970 most sound systems were generally not required to reinforce musical instruments. As a result, often only cell or sectoral type horns were used in sound reinforcement systems because they give excellent results for speech.. These horns come in a variety of sizes and they can be used alone or in multiples to provide tailored arrays to give even sound coverage in rooms. Because of their great directivity characteristics, they will penetrate almost twice as far as any other loudspeaker into a reverberant space with clarity. Drivers called compression drivers that power these horns are supplied separately. You can choose power and tonal characteristics for your horn by the selection of a compression driver from the many available.

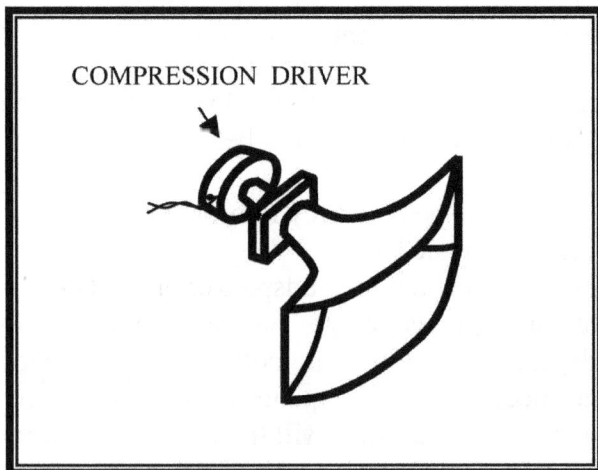

COMPRESSION DRIVER

SECTORAL TYPE HORN
Shown With Compression Driver
Installed

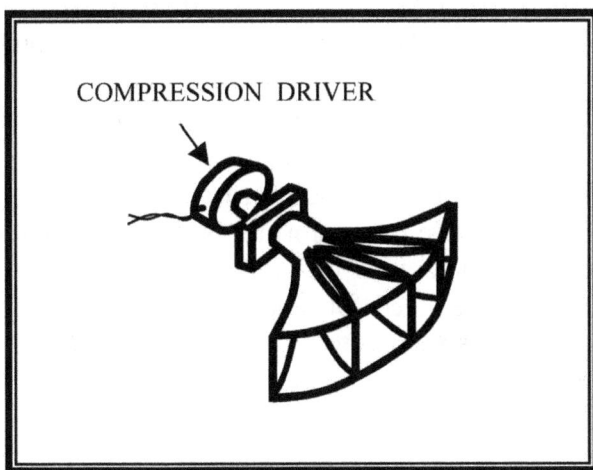

COMPRESSION DRIVER

FOUR CELL HORN
Shown With Compression Driver
Installed

Bass Loudspeakers

When you wish to reinforce musical instruments, horns alone are not adequate. They greatly suppress bass tones below about 150Hz (D3) and must be supplemented with a conventional type loudspeaker or loudspeakers. Paper cone conventional loudspeakers come in a variety of sizes, typically 8, 10, 12, or 15 inch diameters. As a rule, the bigger and heavier, the better low tones are reproduced. While conventional loudspeakers usually have a good overall frequency response and reproduce almost all the tones we hear, they do not have the directivity capability of horns.

In order for conventional loudspeakers to adequately reproduce bass tones, they must be put in boxes called enclosures or baffles. Baffles control the sound from the backside of the loudspeaker that tends to cancel the projected sound from the front side. The size and tonal characteristics of the loudspeaker determine the size of the baffle.

System designers have a wide variety of horns and loudspeakers from which to choose. There are loudspeaker/horn combinations where horns are permanently mounted on or in the baffle providing off-the-shelf arrays that work well in certain situations.

The Ideal Loudspeaker Location

A single location above the performer is the ideal spot to put the loudspeaker array. The most efficient combination for the array is a horn or horns selected for smooth coverage coupled with a loudspeaker or loudspeakers to provide the lower range of tones. Unfortunately, loudspeakers have poor directivity and the horns must be relied on for coverage and clarity.

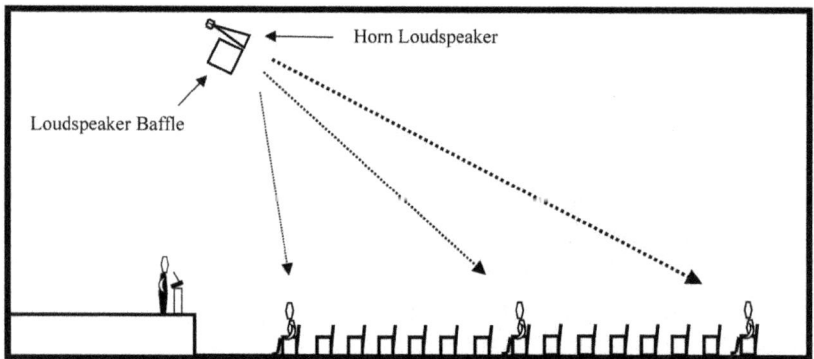

ARRAY POSITION - Sound from the center of the horn is the most intense and it is pointed to the most distant listener. Off center the sound intensity is less, but at the middle and front less is required because of the shorter distances.

Locating the spot for the array is generally simple. Put it as high as possible or practical and a little forward of the performer. Make sure that it is on the centerline of the room. By putting the array high you accomplish three things.

1 – Sound system microphones are as distant as possible reducing the feasibility of sound system feedback.
2 – The array can be focused to the most distant listener.
3 – By directing the projected sound from a high angle, sound energy projected to the back wall is reduced.

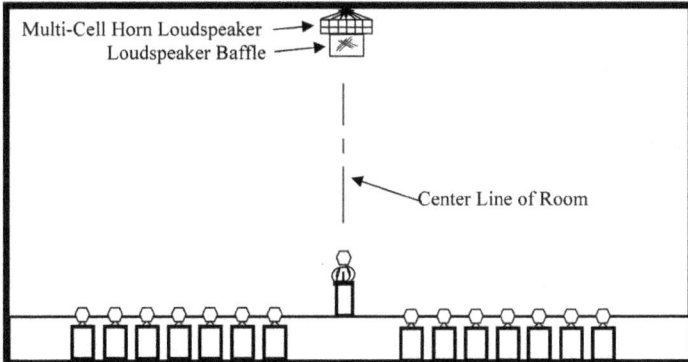

LOUDSPEAKER ON CENTERLINE – When the loudspeaker is on the centerline of the room, the reflections from the projected sound will be symmetrical providing smoother distribution.

Powered Loudspeakers

Powered loudspeakers used as reinforcement or monitor loudspeakers offer advantages over conventional loudspeakers. The most obvious is that you don't need separate external power amplifiers. The power amplifiers are located in the cabinet with the loudspeakers. In a modest-size sound system this alone can eliminate an entire rack of power amplifiers. Another advantage is that the heavy wiring normally required between the external power amplifier and the loudspeaker is eliminated. Small signal wiring similar to microphone wiring is all that is needed between the mixer and powered loudspeaker.

The disadvantages are that the power amplifier volume level

control is situated at the powered loudspeaker and that an AC power outlet is required at the powered loudspeaker location.

Column Loudspeaker Arrays

Another type off-the-shelf array that is used in smaller room applications is the column loudspeaker unit. Here, conventional loudspeakers are placed in a long narrow enclosure that provide in their combination, better directivity. The column unit projects a fan-shape distribution pattern with a wide horizontal and narrow vertical coverage. Note that while column speaker units provide some directivity, they are poor compared to horns.

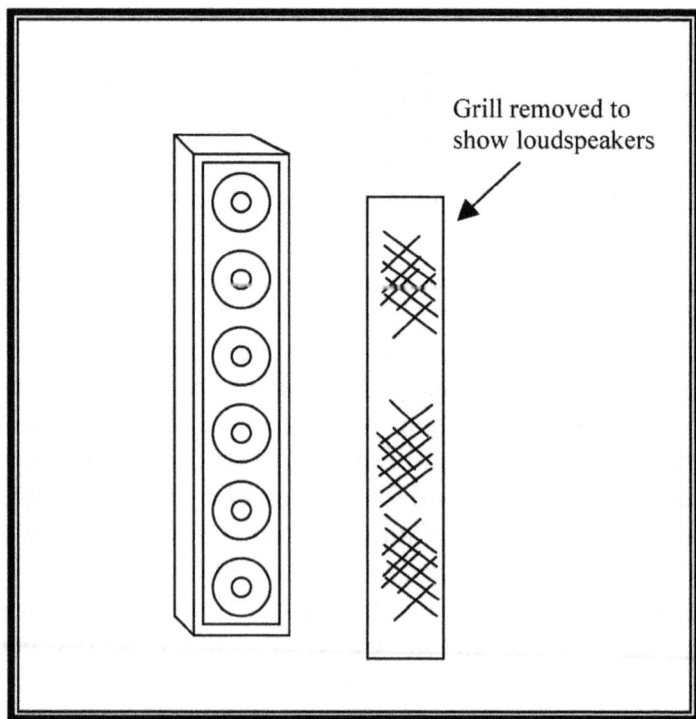

COLUMN LOUDSPEAKER ARRAY

Loudspeaker Location In A Low Ceiling Room

When the single location loudspeaker array is not an option because of room structure, a plan with more than one loudspeaker location must be considered. A room that has a low ceiling is best served by a costly high-density overhead loudspeaker distribution. The proper density is illustrated below. Loudspeakers are designed to have a 90° projection angle and smooth coverage is calculated to the listeners ear level. In this system every listener is not more than 5 to 6 feet from the direct sound of a loudspeaker and listeners this close will not be distracted by any reverberant fields. The birds eye view example of the low ceiling shows that a total of 24 loudspeakers in baffles. are necessary. The cost of quality loudspeakers mounted in proper baffles causes some to skimp by reducing the number of loudspeakers and baffles, eliminating baffles, or using cheap loudspeakers. When this happens, the sound coverage becomes spotty and the sound quality poor.

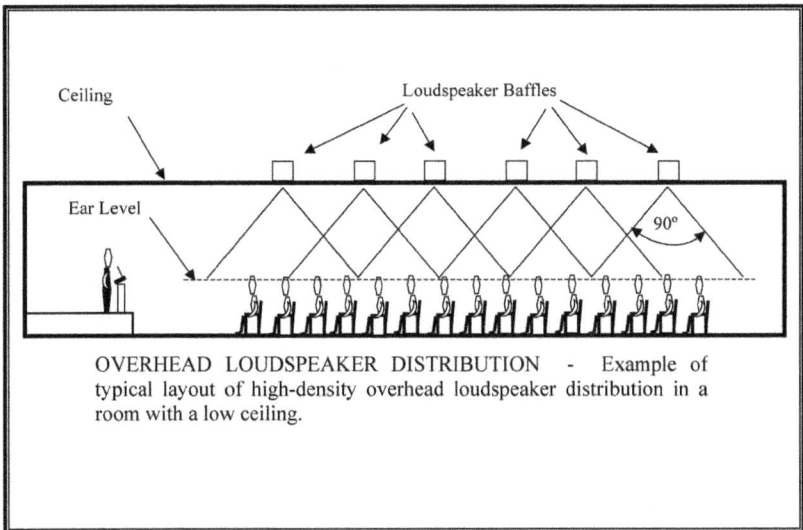

OVERHEAD LOUDSPEAKER DISTRIBUTION - Example of typical layout of high-density overhead loudspeaker distribution in a room with a low ceiling.

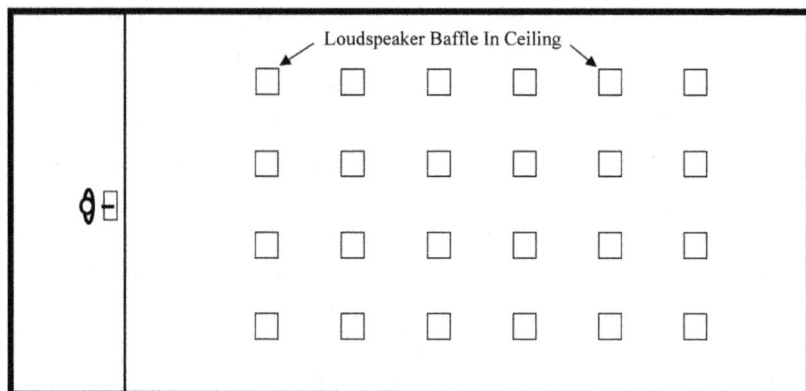

BIRDS EYE VIEW OF LOUDSPEAKER DISTRIBUTION - Example of a typical layout of high-density overhead loudspeaker distribution in a room with a low ceiling.

Improperly Installed Loudspeakers

Installations that result in loudspeakers being mounted low in the room or along the walls should be avoided if possible. Loudspeakers should always be positioned so that sound can be projected to the most distant listener. If they are mounted too low, the near listeners may be overwhelmed and at the same time shadow the sound from the most distant. From each loudspeaker location a reverberant field sets up in the room. Every time another loudspeaker location is added, another reverberant field is created. As locations increase, sound clarity decreases. Furthermore, if the loudspeakers are mounted to the walls, you will see the performance in one place, but hear it in another.

Portable Loudspeaker Systems

Portable sound systems that use loudspeakers on either side of the performer suffer from many disadvantages. Often loudspeakers are not high enough and when they are, they cannot

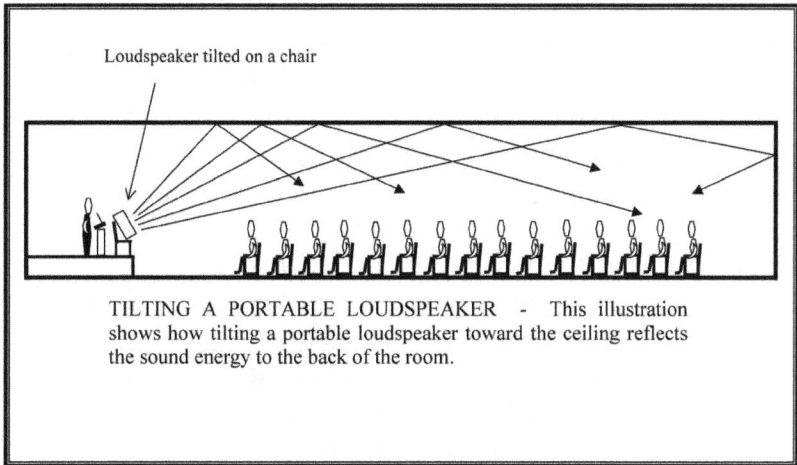

TILTING A PORTABLE LOUDSPEAKER - This illustration shows how tilting a portable loudspeaker toward the ceiling reflects the sound energy to the back of the room.

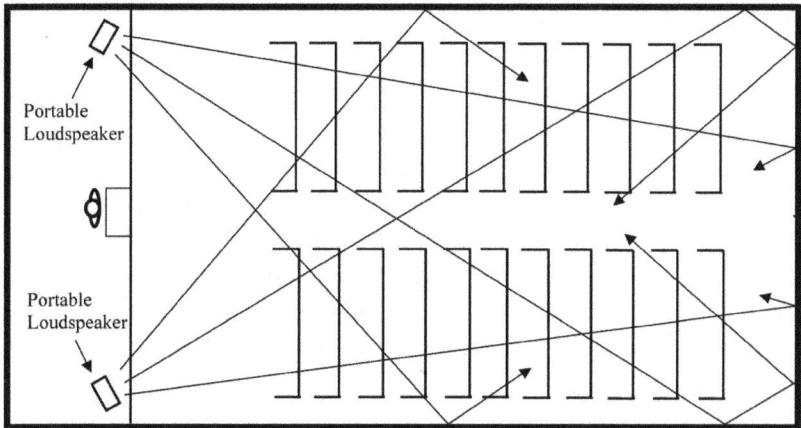

CRISSCROSSING THE PROJECTED SOUND - You may reduce echo, by crisscrossing the projected sound to reflect off the walls.

be directed in a vertical plane because the mountings are not adjustable. Usually the sound is pointed directly into the back wall. Directing the loudspeakers in an unorthodox manner can sometimes help these problems. Tilting them on a chair or table toward the ceiling takes the sound energy that was too loud at the front and reflects it to the rear. Also to reduce echo, you may try crisscrossing the projected sound to reflect off the walls. Remember that sound reflects off smooth surfaces at the same angle it hits and adjusting the loudspeaker angles can be very critical. A little adjustment may affect the results much.

Stereo Sound Reinforcement Systems

When stereo sound mixers became popular and affordable in the 1970's, some sound contractors started selling customers stereo sound reinforcement systems. Stereo design requires dual amplification and dual loudspeaker arrays, and unfortunately, this gimmick to sell more equipment degrades the sound quality rather than enhances it. In order to listen to stereo properly, you must be the same distance from both loudspeakers. This puts the overwhelming number of listeners in the position where the sound is out of balance. Also, the introduction of another loudspeaker array sets up another reverberant field to reduce clarity.

Some reinforcement systems are sold with stereo enhancement to a monaural system. This is where the center loudspeaker array gives total coverage then stereo loudspeakers are added left and right for special effects. This is an expensive addition to a sound system and should be the last thing considered. Since there is such small benefit, money should first be spent for needed quality primary equipment.

Monitor Loudspeakers

Monitor Loudspeakers in a sound reinforcement system can sometimes present problems if certain details are not observed. Consider the following suggestions:

1. Select the right type monitor loudspeaker for the situation. How much area is the monitor to cover? Is a floor "wedge

loudspeaker" the right choice or is a small stand mounted loudspeaker the best?

2. Try to place the monitor loudspeaker at the least sensitive point of a microphone's pickup.
3. Use the minimum number of monitor loudspeakers.
4. If monitor loudspeakers are so loud that they affect what the congregation hears, turn down the volume or relocate the loudspeakers and/or the performers.
5. When you incur certain instruments that are too loud causing singers or other instrumentalists to require loud monitor loudspeaker volume, experiment with repositioning the performers to minimize the problem.
6. If you have any instrument amplifiers on the stage, place them in front of and pointing toward the musicians, and away from the congregation and other performers.

One of the best ways to eliminate the above problems is to employ in-the-ear monitors which can be wired or wireless. These tiny monitors fit into the ear and are practically invisible to the congregation. Since there are no monitor loudspeakers, each performer can set his monitor listening level without interfering with someone else. When monitor loudspeakers are eliminated there will be less chance for sound system feedback and less chance of disturbing the congregational sound balance.

Choirs occasionally need monitor loudspeaker coverage to be able to sing with recorded music tracks. The best solution here is to provide as many monitor loudspeakers as needed as close to the choir as possible. The difficulty here is if the overhead reinforcement microphones are used, a sound balance to the congregation may be hard to achieve. If the choir needs monitor loudspeakers for on stage live music, you may find that the sound feedback possibility is also heighten. Not using the overhead reinforcement microphones will help some.

Part Three
SOUND SYSTEM MICROPHONES

REINFORCEMENT MICROPHONES

Microphones detect sound that will be processed, amplified, and then distributed through loudspeakers to selected listeners. A good sound reinforcement system is no better than the microphone selection and the microphone placement techniques used. The skill exercised here will determine how well a good sound system performs.

All microphones in any good sound system must have a balanced low impedance signal output feeding a balanced low impedance microphone input into the sound system. If the microphones are not balanced or the sound system equipment is not balanced, the microphone lines becomes susceptible to noise and radio signals. To assure proper operation, the microphones must be properly connected to the sound system with quality cables and connectors.

Dynamic and condenser microphones are most commonly used in the typical sound reinforcement system. Dynamic microphones are rugged and economical, and provide good performance just about anywhere they are utilized. Condenser microphones, sometimes called "capacitor microphones", tend to be more delicate and do not perform well where the sound is very loud and where close microphone techniques are used. Condenser microphones provide a crisp and bright output while dynamic microphones tend to be mellower.

Dynamic Microphones

Before the popularity of condenser microphones, dynamic microphones were used in almost every application. These work well as hand held microphones and where close microphone techniques are used. Since they work well in loud environments, they are ideal for used with loud musical instrument amplifiers. When it became common to sing loudly or shout into the microphone one to two inches from the mouth, vocal performance dynamic microphones were developed to handle the high pressures. Note that precaution must be taken to prevent sound system distortion overload when this is done. The input attenuator adjustment on your sound equipment mixer will reduce the large signal level from the microphone.

Condenser Microphones

Almost all condenser microphones used today are "electret condenser" microphones that must be powered by a battery or an external power supply. Most sound system microphone mixers provide that external power, called "phantom power", through the microphone cables to eliminate the need for a battery or any other power source. Condenser microphones cannot handle very loud sound and they are best suited for the podium, lavaliere, choir coverage, acoustical instruments, etc. They generally have higher sensitivity than dynamic microphones and provide a smoother, more natural sound, particularly at high frequencies. Condenser microphones can be made very small without compromising performance. Because they are more delicate, temperature extremes and humidity may cause them to fail temporarily. Condenser microphones are more complex than dynamic microphones and they tend to be more costly. Condenser microphones that are not the electret type cannot be powered by phantom power from the sound equipment and must have their own external power supply.

Boundary Microphones
Boundary microphones were designed to be mounted on surfaces such as a piano lid, wall, stage floor, table, or panel. They are miniature electret condenser microphones configured to reduce or eliminate the effect caused by reflections off nearby hard surfaces that create two or more sound paths to a distant microphone. When properly mounted, they eliminate the hollow sound that would result by letting reflected sound combine with the direct sound. See illustrations on page fifty seven.

Microphone Basics
The choice of microphone for a given application is subjective. There is no ideal microphone or ideal microphone placement technique and these choices are ultimately the result of the operator's judgment. Experimenting with different microphones and different placements will help you achieve the best result. Remember that microphones do not selectively sense sound like our ears; they will pick up all the sound in the area.

Always get the microphone as far away as you can from any sound reinforcement or monitor loudspeaker and place it as close to the sound source as practical. This will minimize the possibility of sound system feedback. The closer the microphone is to the sound source, the less amplification the sound system will need to provide. Each time the distance between the microphone and the sound source is reduced, the sound level signal from the microphone will increase markedly. Avoid using more microphones than necessary and turn off connected microphones that are not being used. Every time a microphone is added to the sound system, the risk for sound feedback increases and useable sound system amplification is reduced.

Omnidirectional and Unidirectional Microphones Compared

Both dynamic and condenser microphones are made in omnidirectional and unidirectional configurations. An omnidirectional microphone picks up sound in all directions at nearly the same sensitivity. Unidirectional microphones have greater sensitivity to sound pickup in one direction.

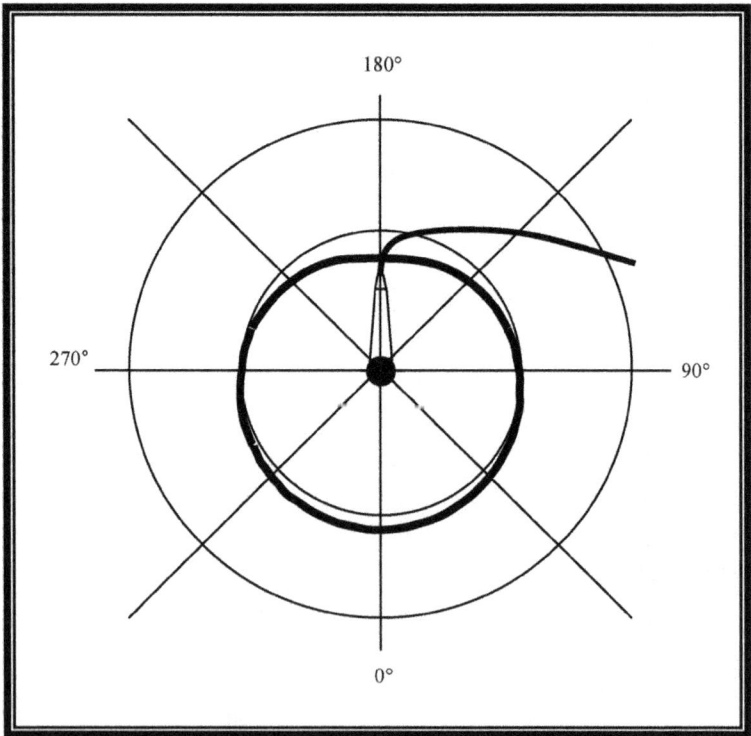

OMNIDIRECTIONAL MICROPHONE POLAR PATTERN
Simplified typical omnidirectional microphone polar pattern chart. Notice that the pattern shows a near even pickup around the microphone. Sound pickup is slightly less on the sides and a little less at the rear.

The manufacture provides data for microphones that illustrate the pickup pattern for each microphone. These are called "polar pattern" charts. At the center of these charts where all the degree lines converge is the sound pickup diaphragm of the microphone that is pointed toward 0°. In the polar chart illustrations the chart is superimposed on the microphone for clarity. The pattern shows how close a constant level tone would need to be at any point around the microphone to provide a constant level signal output.

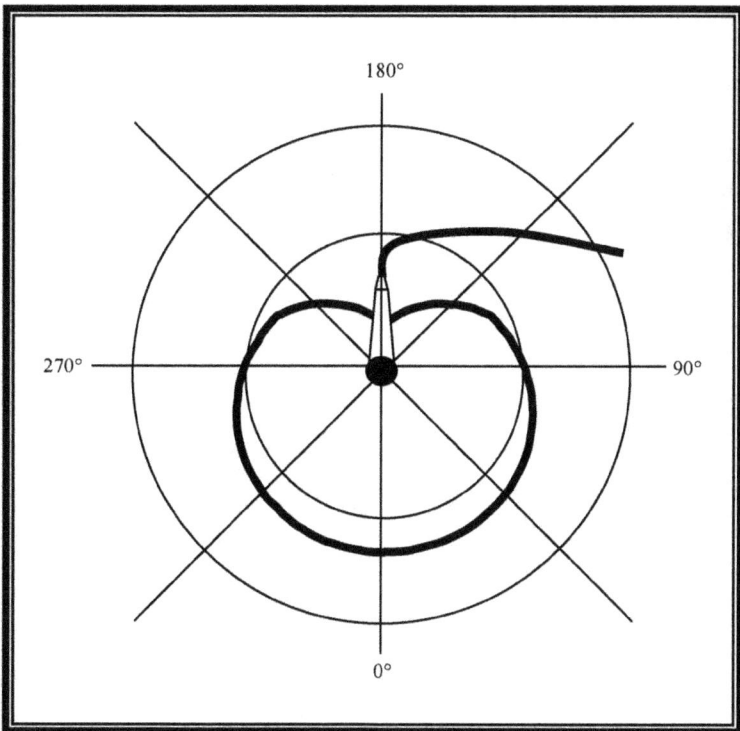

CARDIOID MICROPHONE POLAR PATTERN
Birds eye view of a typical unidirectional (cardioid) microphone polar pattern chart. Notice that on this simplified chart that the best sound pickup is directly in front of the microphone and the poorest is at the rear.

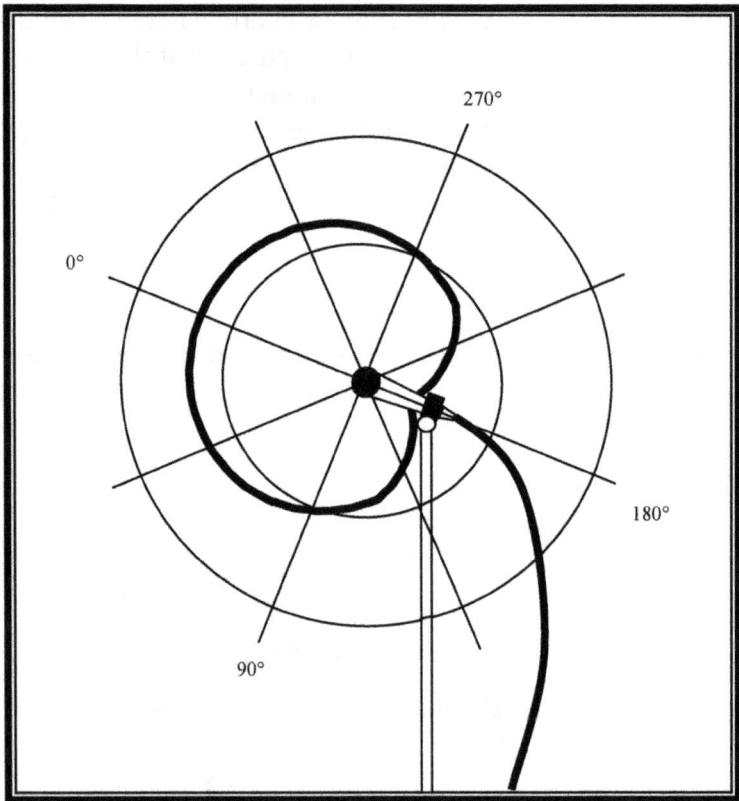

POLAR PATTERN CHART SUPERIMPOSED ON A MICROPHONE Typical simplified unidirectional (cardioid) microphone polar pattern chart superimposed on a microphone mounted on a stand. Notice that the best sound pickup is directly in front of the microphone and the poorest is at the rear.

As you can see, the omnidirectional microphone pickup pattern is close to the same distance front and back, and side to side showing that sound is picked up near equally from every direction. But, the unidirectional microphone pattern shows that the constant level tone must be closer at the sides and back of the microphone to provide a constant level signal output.

Since the sound must be closer at the sides and back, the sound at the front of the microphone will be picked up better. The polar pattern for unidirectional microphones is heart shaped. As a result, these one-direction microphones are often call "cardioid microphones".

Unidirectional or Cardioid Microphones

While unidirectional microphones help to reduce the pickup of unintended instruments or other sounds, and help to lessen sound system feedback, they suffer from "off axis coloration" and "proximity effect". When sound is not project directly straight into the front of the unidirectional microphone, off axis coloration takes place. The more off center you are, the more high tones are reduced.

Proximity effect is a bass boost that takes place when the sound source is closer than two feet from the microphone. The closer to the microphone is to the sound, the greater the bass boost that may create a "muddiness" in the sound. Many unidirectional microphones have compensating "bass roll off" switches that can be switched "on" when the sound source is closer than two feet.

Because unidirectional microphones pick up less unwanted surrounding sound than omnidirectional microphones, they may be used at greater distances from their sound source and still achieve the same balance between the direct sound and unwanted sound.

Here, we discuss the use of cardioid microphones because they are more common than supercardioid and hypercardioid microphones that have greater pickup directivity and higher susceptibility to off axis coloration. If you have supercardioid and hypercardioid microphones, study their polar pattern charts to determine how best to use them.

Omnidirectional Microphones

Omnidirectional microphones have equal pickup of sound in all directions, i.e. up and down, side to side, and back and forth. These smoother response type microphones are often preferred when feedback control or suppression of surrounding sound is not critical. Omnidirectional microphones do not suffer from "off axis coloration", "proximity effect" or "breath pop sensitivity" that bother unidirectional microphones. Because omnidirectional microphones are not directional, they cannot suppress unwanted sounds from other sources. They should be placed as close as possible to their sound source to pick up a useable balance between direct sound and unwanted sound.

Evaluate the Sound Source Environment

Examine the acoustical environment. Look in all directions including up and down to determine the acoustical condition. Are there any reflective hard surfaces nearby? Are there any building mechanical noises, light fixtures buzzing, or high volume air flows?

When a microphone is located some distance from the direct sound and a hard surface is nearby, the direct sound will also reflect off that surface. Reflected sound travels a longer path than the direct sound, so the reflected sound is delayed relative to the direct sound. The direct and delayed sounds combine at the microphone. This results in a "lack of presents" or hollowness in the sound picked up. If you cannot remove the offending hard surface, move the microphone closer to the sound source and/or use a properly positioned directional microphone to reduce the problem. Place the microphone near the floor when a hard floor is the problem. You may find an omnidirectional microphone works best here on a floor mount designed for this purpose or else use a boundary microphone on the floor. This will eliminate the reflected sound pickup, but direct noise near the floor will be more easily picked up. While this floor technique works, you should otherwise avoid placing a microphone near a highly reflective surface if possible. Or else investigate using a boundary microphone.

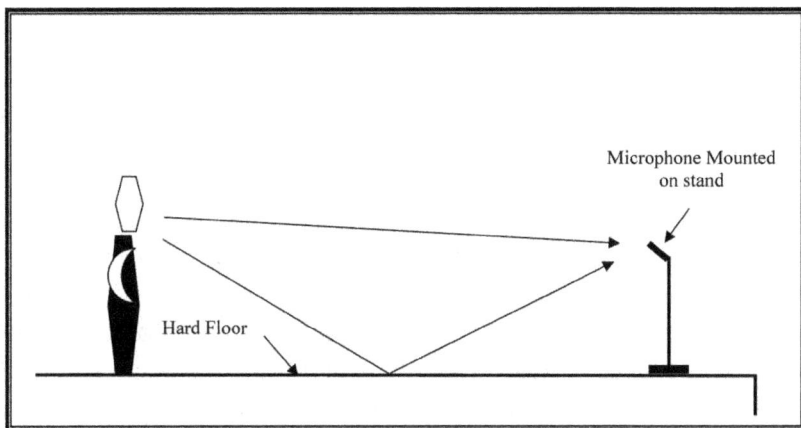

MICROPHONE DIRECT AND REFLECTED SOUND - When the microphone receives from the source a direct sound and a strong reflected sound, the resultant sound picked up is hollow.

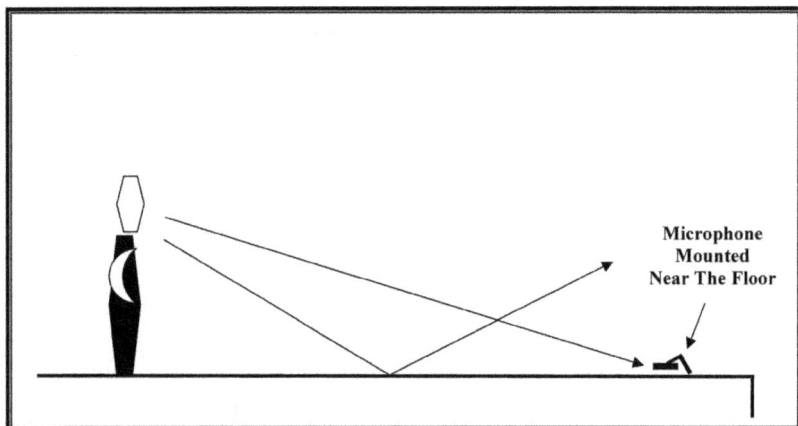

FLOOR MOUNTED MICROPHONE - When the microphone is placed near the floor or a boundary microphone is employed, the reflected sound is not picked up.

Beware of airflow that can create noise passing over the microphone. This may occur with microphones suspended from the ceiling where high volume airflow is most often found.

Study Your Sound Sources

Determine the nature of the sound source. Is it speech, singing, musical instruments, etc. Is it soft, loud or, somewhere in between? Study the sound radiating characteristics of your sound sources to determine the best microphone pickup technique.

Speakers or singers who do not project their voice well and musicians who play tentatively are hard to pickup because they produce an inconsistent sound level. Generally the best technique here is to get the microphone as close as possible and hope for the best. Unfortunately a sound system will not often make a poor sound source sound better. More often than not, the sound system will accent the bad in a performance.

Ten Foot Rule

Only one microphone should be used to pickup one sound source when possible. If two or more are used to pick up one sound source, they should be at least ten feet apart. The microphones should have similar pickup characteristics to keep the sound uniform. Using microphones that are alike is best. The most common place you will find multiple microphones covering a single sound source is a choir.

Microphone Positioning

Microphones must often be located in positions where they may pick up unintended instruments or other sounds. Whenever a single source is picked up by more than one microphone, the sound arrives to each microphone at different times. The direct sound and the delayed sound mix in the sound equipment to produce a resulting "lack of presents" or hollowness to the sound. A monitor loudspeaker improperly placed where a microphone picks up its sound will produce the same effect. This is where you use your unidirectional microphones to an advantage. Find the right microphone position to maximize the desired sound and minimize the undesired sound. Look at the polar pattern chart on your microphone to find that position. Aim the monitor loudspeaker toward the least sensitive spot on the unidirectional microphone.

3 to 1 Rule

When there are multiple sound sources requiring multiple microphones located near one another, the 3 to 1 rule should be observed. If a sound source is one foot from its microphone, the next nearest microphone should be at least 3 feet away. If the sound source is two feet from the microphone, the next nearest microphone should be at least 6 feet away, etc. By observing this rule, you will minimize interference problems between microphones.

Mechanically Noisy Microphones

No microphone is totally immune to handling noise although some designs are very good. In situations where you have excessive microphone pickup through stands or other mounting devices, you can use one of the many accessory shock mounts that are available. Note also that noisy microphones are more susceptible to causing sound feedback.

To reduce "breath pops" when speaking close to a microphone, use a microphone with a built in breath pop filter or place an external foam windscreen on the microphone. A microphone tends to be most sensitive to breath pops at about 3 inches. You can move closer or farther away to reduce the pops. Consider using an omnidirectional microphone. They are less susceptible to breath pops.

Summary Of Things To Consider
When Placing A Microphone

1 – Evaluate the surroundings. Are there hard surfaces, mechanical noises or high airflow that you need to move away from, suppress, or eliminate?

2 – Study your sound source radiating characteristics. Do you want a dynamic microphone or would a condenser microphone be better? Remember that a condenser microphone will need to be powered.

3 – Do you need a unidirectional microphone to suppress surrounding unwanted sounds or reduce sound system feedback possibility? Is an omnidirectional microphone right for this application?

4 – Place the microphone as far away as possible from the reinforcement and monitor loudspeakers.

5 – Put the microphone as near to the sound source as possible and aim the microphone toward that source.

6 – When using a unidirectional microphone, pay attention to off axis coloration and proximity effect.
    Note that off axis coloration may be used to your advantage if the sound source is harsh or sharp sounding.

7 – Position a unidirectional microphone to best reject unwanted sounds. Know the polar pattern response for the microphone.

8 – Do not use more microphones than necessary to cover a single sound source. If you use more than one, observe the 10-foot rule.

9 – Note the nearby microphones and observe the 3 to 1 rule.

10 - Determine if there is sufficient mechanical noise control. Thumping on a lectern or walking on a bouncy floor is often transmitted mechanically through the microphone stand or gooseneck. You may need to use an accessory shock mount. If the microphone has a bass roll of filter switch, switching "on" the filter may help.

11 - Use an external foam windscreen on your unidirectional microphone or employ an omnidirectional microphone to reduce breath popping

12 - Clip a lapel microphone about 6 inches below your chin and route the microphone wire so that the microphone is not tugged or rustled. A headset microphone provides the closest microphone to mouth arrangement by placing the microphone at the corner of the mouth. This may be the preferred microphone if the person speaks softly or if sound system feedback is a problem. These microphones are always used in wireless configurations.

Covering a Choir

It is impossible to get a high level of sound reinforcement for a choir in the typical church or auditorium. The use of multiple microphones 3 to 4 feet away from the source presents a need for so much sound system amplification, that the sound system will feedback. As a result, a very low level of sound reinforcement is all that is possible. Although all churches seem to think that their choir must have sound reinforcement, I do not know of one choir microphone coverage that works well. Before the sound amplification level is turned up sufficiently to actually make a difference, sound system feedback occurs.

I have seen several approaches for placing suspended microphones above the choir and usually cardioid microphones are hung high above the first row and pointed toward the back row. When microphones are placed too close to the choir, stronger voices are accented causing an imbalance to the overall sound of the choir. When they are too far away, the sound pick up is not enough before sound system feedback happens. It is not possible to make a 20 member choir sound like a choir of a hundred.

Microphones placed for choir sound reinforcement do not provide good placement for recording. A recording device will receive only the sound from the microphones while those in the auditorium hear a balance comprised of acoustical sound plus the sound picked up by the microphones. Recording microphone techniques are discussed in the next section.

Rules for microphone positioning are often broken when choir coverage is attempted. The two-cardioid microphone placement is usual for as many as 30 voices. Space the two microphones at least 10 feet apart and about 10 feet above floor level of the first row. Divide the choir into two sections with an imaginary line, then suspend a microphone at the front and center of each division and point it to the last row. If the width of the choir is as much as 30 feet wide, another microphone could be added in line with the other two. Condenser microphones work best in this application. Typical microphone placement is shown in the illustrations.

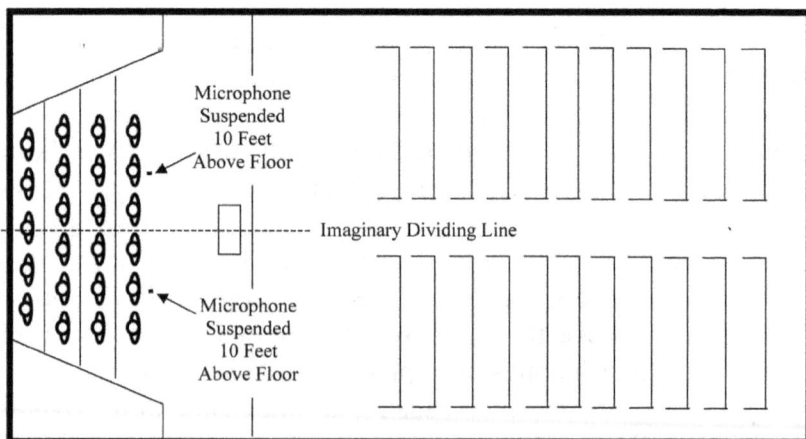

BIRDS EYE VIEW OF CHOIR REINFORCEMENT MICROPHONES - Birds eye view shows the choir divided by an imaginary line. Suspend cardioid microphones at the front and center of each division and point them to the last row. Try to keep the microphones at least 10 feet apart.

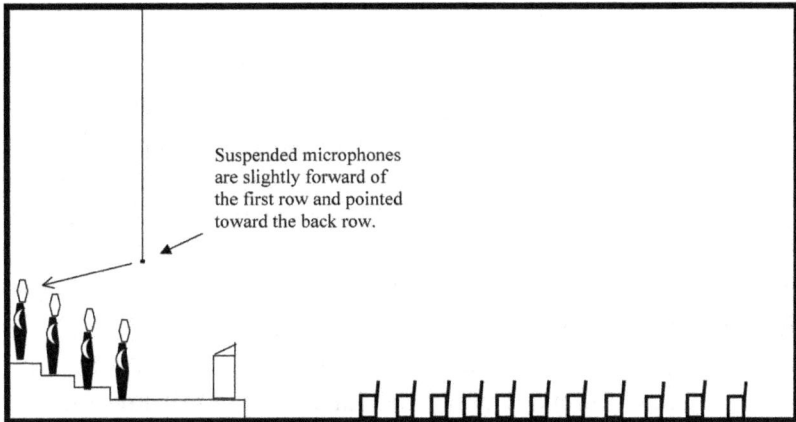

SIDE VIEW OF CHOIR REINFORCEMENT MICROPHONES - Side view shows cardioid microphones suspend ten feet from the floor. They are slightly forward of the front row pointed toward the last row.

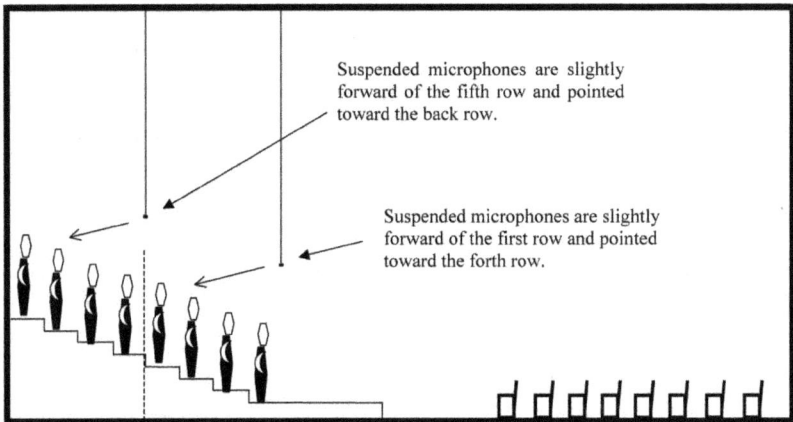

SIDE VIEW OF DEEP CHOIR REINFORCEMENT MICROPHONES - Side view shows an unusually deep choir divided by an imaginary line. Suspend the cardioid microphones at the back division in the same manner as the first. Mount the microphones 10 feet above the floor and try to keep them at least 10 feet apart.

Fortunately, most churches have choir areas that have hard walls and ceiling that provide natural acoustical reinforcement making the choir quite listenable without sound system help.

Choir Stereo Recording Microphone Techniques
If you use sound reinforcement microphones to record the choir or feed a broadcast, a poor recording or broadcast will result. The sound picked up by close microphones for sound reinforcement is acceptable for most people because the acoustical sound of the choir dominates. If you listen only to the sound picked up by the microphones, you will hear spotty and uneven sound. The ideal spot to place a microphone for choir recording is far enough to get an acoustical blend, but not too far where words are not heard clearly. Cardioid condenser microphones work best for this application.

The illustration in the following shows a two-microphone stereo placement for a choir. The forty-degree angle goes to the edges

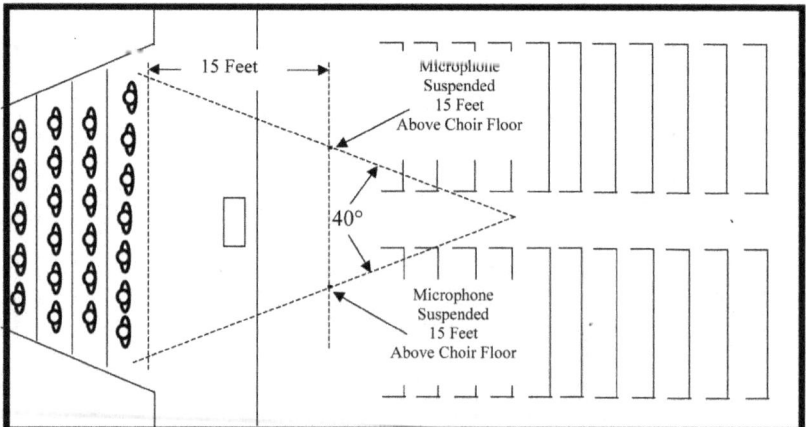

BIRDS EYE VIEW OF CHOIR RECORDING MICROPHONES - Birds eye view shows the choir microphone placement for recording. Suspend cardioid microphones 15 feet above the choir floor and 15 feet from choir front row where the 40° angle intersects. Point microphones to back row.

of the front row of the choir to establish the distance between microphones. Notice that the distance from the front row of the choir and the height of the microphones remains at 15 feet even though the choir size changes. Point microphones toward the back row.

The next illustration shows the microphones in the X-Y stereo configuration. Here, two identical cardioid microphones are used with their capsules one above the other as close as possible and setting at ninety degrees to one another. The X-Y Pattern provides good stereo separation and is often preferred because it gives excellent mono compatibility. Suspend the microphones in the X-Y pattern 15 feet in front of the first row and 15 feet above the choir floor.

The X-Y pattern also works well for stereo recording large bands, pipe organs, and orchestras.

CHOIR

Identical cardioid microphones with their capsules one above the other as close as possible

Microphones set at ninety degrees to one another

X – Y Stereo Microphone Pattern

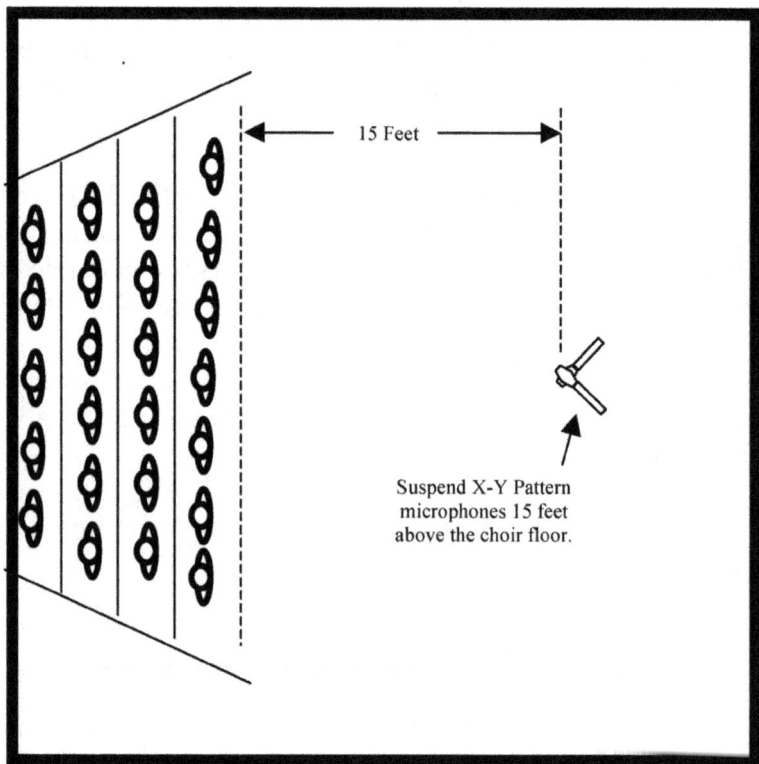

X-Y STEREO MICROPHONE PATTERN PLACEMENT - Birds eye view shows the choir X-Y stereo microphone pattern placement for recording. Suspend cardioid microphones 15 feet above the choir floor and 15 feet from choir front row. Point microphones to back row.

Stereo microphones are made that you may use instead. Place the stereo microphone at the same height and distance you suspend the X–Y Pattern microphones.

Microphone Pickup of Drums

Acoustic drum sounds are often difficult to contain. The sharp staccato sounds easily project and bleed into other nearby microphones creating multiple path pickups that result in drum sounds that lack "presents" and sound hollow. When drums must be jammed close to other instruments or an amateur drummer does not know how to control the loudness of the drums, the problem increases. The easiest solution is to use electronic drums or to acoustically isolate the acoustic drums. By enclosing the acoustic drums, you put a barrier between the drums sounds and adjacent microphones meant for other sound sources. Do not forget to isolate from any suspended microphones hanging from the ceiling. Plexiglas enclosures are popular, but drum sounds reflect off the rigid material inside the enclosure. An absorbent panel must be placed behind the drummer to suppress these reflections.

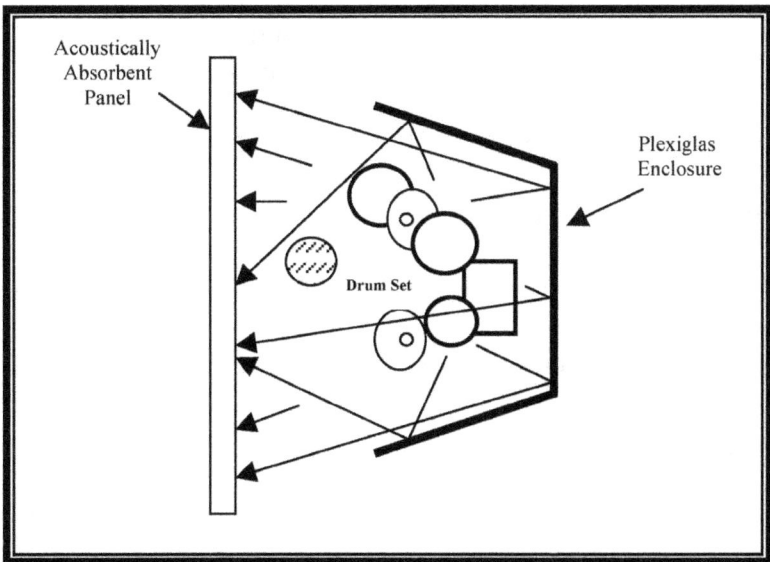

DRUMS IN A PLEXIGLAS ENCLOSURE - Birds eye view of drums in a Plexiglas enclosure showing how sound is reflected into a sound absorbent panel to prevent drum sound from bleeding into adjacent microphones.

Three microphones will usually adequately cover a drum set. Place the first one a few inches in front of the bass drum, place the second microphone one inch from the rim of the snare drum, and suspend the third two to three feet above the set to pick up the tom-tom drums and symbols.

Microphone Techniques for the Performer

1 – When testing a microphone, do not blow into it. The operator cannot determine the sound system adjustments until you speak or play into the microphone.

2 – Hold microphone at proper distance from your mouth. Avoid the breath popping distance of about 3 inches. Hold it a little closer or a little farther back. Do not hold the microphone very close to your mouth if you cannot maintain a constant voice level. Hold the microphone at a proper distance for balanced sound.

3 – Aim the microphone toward your mouth and away from other sound sources.

4 – Use a microphone wind screen to control popping if necessary.

5 – Switch the microphone bass roll off switch "on" to compensate for the proximity effect that occurs in unidirectional microphones if you are closer than two feet.

6 – Do not create noise by excessive handling of the microphone and cable.

7 - Control dynamics with your voice rather than moving the microphone.

8 –When you use a wireless lapel microphone, clip it about 6 inches below your chin and route the microphone wire so that the microphone is not tugged or rustled.

9 – Wireless headset microphones provide the closest microphone to mouth arrangements and may be a better choice if you speak softly or if sound system feedback is a problem. Headset microphones have a lightweight frame that sets on the ear and wraps over or around the

back of the head and places the microphone pickup at the corner of the mouth.

10 - Speak in a clear and distinct voice.

11 – If you are holding and playing an instrument, maintain your position relative to the microphone.

Wireless Microphones

A wireless microphone is a battery operated microphone combined with a radio transmitter that sends signals through the air to a radio receiver located at the sound system equipment. Wireless microphones have tiny FM radio transmitters made to operate on VHF (very high frequencies) and UHF (ultra high frequencies). VHF wireless microphones operate on the same frequencies as VHF TV stations and should be selected with frequencies of TV channels not operating in the area. The same is true of UHF microphones. They share the UHF TV band. VHF and UHF transmitting frequencies are often called RF (radio frequencies), named before the invention of television.

A wireless microphone receiver can receive only one microphone radio signal. Each wireless microphone must have its own receiver that is "tuned" to its radio frequency. VHF microphones cost less than UHF microphones but they are more susceptible to interference from machinery and computers. There is less potential TV station interference with UHF microphones because UHF stations are generally located on the outskirts of town and operate with lower transmitting power. When purchasing wireless microphones, always ask the vendor to provide microphones at frequencies least likely to be interfered with by TV stations in the area. Also when buying more than one microphone, ask to provide frequencies that are least likely to interfere with one another. Less expensive wireless microphones do not have the built-in designs that prevent certain frequency microphone transmitters from sometimes clashing.

A straight wire or a coiled "rubber duck" antenna is normally supplied with a wireless microphone. The wire antenna is at the proper length to provide the greatest efficiency and must be kept straight. The coiled antenna was designed to provide that efficiency with a shorter length, but it is sensitive to close proximity conductors such as metal and the human body. If a coiled antenna must be used on a bodypack transmitter, bend it out slightly so that it does not rest against the performer.

The Battery life for wireless microphones varies from model to model. Alkaline batteries provide better and longer performance than ordinary carbon-zinc batteries. The more expensive lithium batteries work well and will last three times longer than alkaline batteries. If you choose to use rechargeable batteries, use alkaline rechargeable batteries. Rechargeable NiCad and NiMH do not work as well in this application.

The most common wireless microphone receivers have an antenna or antennas that are permanently attached. They should be located on a table or shelf in sight of the wireless microphones being used and away from interference causing equipment such as CD players and lighting controls. Place VHF receiver antennas at least 16 inches from any other VHF receiver antennas and UHF receiver antennas at least 4 inches from any other UHF receiver antennas to prevent lowering the efficiency of reception.

When walls, large metal objects, and many people are between the wireless microphone receiver and its associated wireless microphone, transmission may be interrupted. Be careful of receivers that are mounted in racks or cabinets located out of the line of sight unless they have detachable antennas that can be properly located.

The useable distance between wireless microphones and their

associated receivers varies from model to model and effort should be made to make the distance as short as possible. When there is a long distance between the stage and sound equipment, receivers are often located just off stage and an audio cable is run from the wireless receiver to the sound system equipment.

Today's wireless microphones are much better than wireless microphones made 20 years ago and new innovations are making them more efficient while allowing greater numbers to be used together.

Part Four
SOUND SYSTEM ELECTRONICS

SOUND SYSTEM CONSOLES

Sound System Consoles have many names. Some of these are boards, sound boards, mixing boards, audio boards, mixing consoles, sound mixers, mixers, etc. Most often today they are just called mixers. Compact versions of the mixer containing power amplifiers to power loudspeakers may be purchased, but they are limited in their application and flexibility.

Computer-based systems have been developed that configure sophisticated reinforcement sound systems, but in almost all applications, the basic mixer does the job nicely. The modern mixer was developed over many years by the broadcast and recording industries and has been adapted for use in sound reinforcement systems. That is the main reason that stereo mixers are found in monaural sound reinforcement systems.

Mixer Location
It is necessary for the operator of a sound reinforcement system to hear what the audience hears. He or she must be able to balance and blend the electronically boosted sound with the acoustic sound. This cannot be accomplished with headphones or in a booth through an open window. Neither can this be accomplished in an alcove, or in a corner of an auditorium, or a

sanctuary. Unfortunately, the prime spot for a mixer to be located is about two thirds back and in the center of the auditorium or sanctuary where the audience may be disturbed and the equipment is not secure. An open-air box seat arrangement is the usual design for the mixer and operator that will provide a little separation from the audience. Another place the mixer may satisfactorily be located is front and center of an open conventional balcony. This position is less intrusive to the audience.

Placing the mixer and operator to the side near the wall, or too far back and near the back wall, makes it very difficult to balance and blend the sound and it is difficult for the operator to know when the sound is actually properly adjusted.

The Operator
Sound system equipment is expensive and few churches can afford the highest quality. Many, if not most churches have poor acoustics and most volunteer sound operators have little or no training. Therefore, it is easy to understand why church people often complain about the sound. Away from church the congregation listens to professionally produced CDs, tapes, and other media, and then expect the same quality in their own church. For churches to achieve this, the acoustic environment must be favorable, the sound equipment must be properly installed, and the operator must be knowledgeable and experienced.

When a sound reinforcement system is properly installed and operated, it should not draw attention to itself. It is meant to capture a live presentation and augment it without restricting a performer's concentration or expression. Whenever a sound system or its operation impairs the performance, the sound system and operator have lost their usefulness. Sound operators must be aware of this fact. Sound volume levels that are too high or low, unbalanced audio mixes, monitors that cannot be heard by the performers, and sound system feedback are some of the common problems that can ruin the atmosphere and upset the composure of the performers.

The sound operator's job is important and must be taken seriously. He or she should have a sense of what quality sound is, and should know when sound is balanced. He or she should be knowledgeable in the overall sound system operation and know its abilities and limitations by being familiar with the sound equipment operation manuals. An operator must be willing to take directions and cooperate with performers and other technical staff.

A sound system operator develops his skill by taking advantage of choir, ensemble, orchestra, and drama group rehearsal times. This is the time when new microphone techniques can be tried, features and setups can be experimented with on the mixer, and when technical problems can be solved.

An operator who is touchy, easily offended, arrogant, unfaithful, unwilling to share technical knowledge, or whose ego is fed by his operator status, is a person who should not be an operator.

Operation of the sound system is often a thankless job. When you do a good job, you are not noticed and when something goes wrong, you are embarrassed before the whole crowd.

Mixer Basics
The principal function of the mixer is to route the input signals to where they are desired and to provide moment-by-moment control that those signals demand. Input control options and signal combining are two steps provided by the mixer. Before the input signals reach the combining section of the console, they may be processed in terms of loudness, filtering, equalization,

INPUT CHANNEL

MIKE
INPUT

TRIM

HI PASS
FILTER

EQ

INSERT
POINT

MUTE

PRE-FADE
AUX

PRE-FADE
AUX BUS

FADER

POST-FADE
AUX

POST-FADE
AUX BUS

PAN

ROUTING
SWITCH

SUB GROUP
FADER

MIXER
FADER

MIXER
OUTPUT

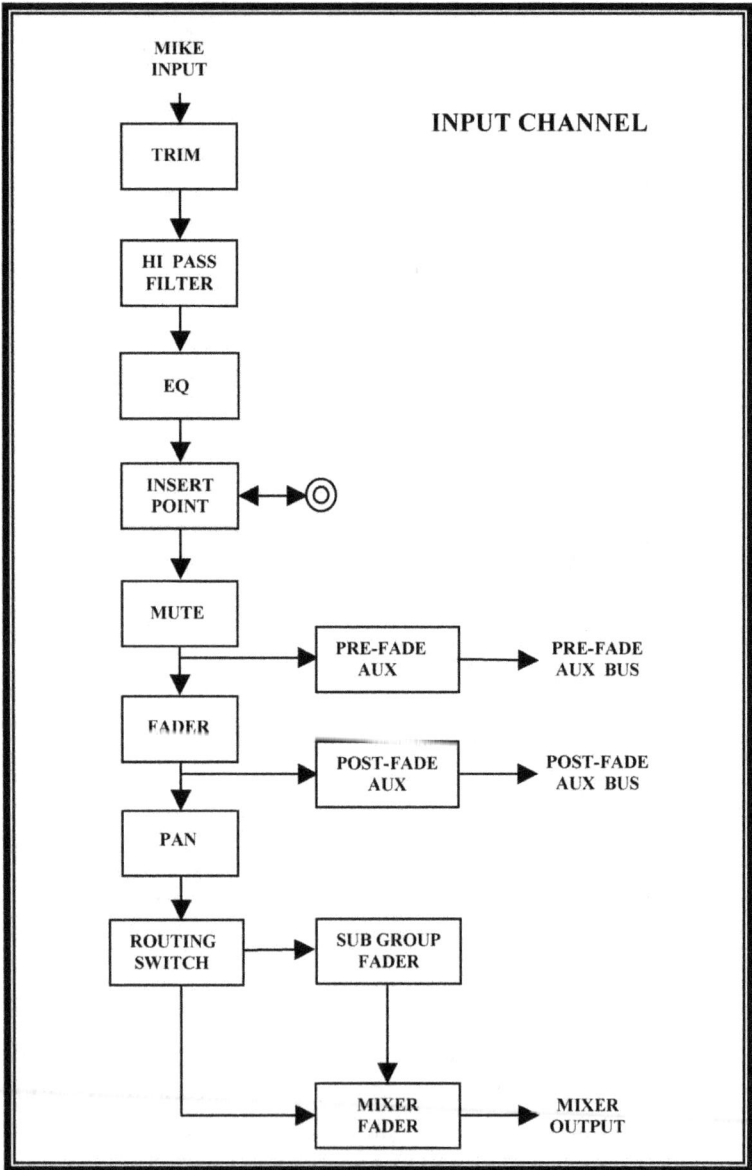

SIGNAL FLOW PATH THROUGH AN INPUT CHANNEL - Simplified Block
Diagram showing a typical signal flow path through an Input Channel.

signal processing, and muting.

The heart of a mixer is the combining section or network. The combining network lets you route input signals to one or more outputs. A well-designed combining network will allow combining any number of inputs to any of the outputs or output buses. A bus is a collecting point for signals that you want to combine and send together before they enter a loudness or level control. A mixer with 16 inputs and 4 outputs is able to send 0 to 16 inputs to any or all of the 4 outputs.

Another combining network used in mixers is the auxiliary network. This network provides signal sends to processing units such as electronic reverberation and to stage monitoring systems. Small mixers may have only two auxiliary sends per input channel module, but larger may have as many as eight.

All mixers generally work the same way although each manufacture designs the control schemes differently. Also the number of input channels, mixer outputs, auxiliary sends, level meters and minor features vary widely. Manufactures use block diagrams to show the routes signal sources take through the mixer. These diagrams provide a visual description of the key elements of the mixer and they should be studied in order to know the mixer's capabilities.

Mixer Input Channels

The first function of a mixer is to process each individual sound signal that arrives to an input of the mixer. Each mixer input has its own dedicated set of processing controls to adjust the volume loudness and tonal qualities of the input source. A mixer input and its set of processing controls constitute a channel. A 16-channel mixer will accommodate a maximum of 16 different source sound signals. A channel may also be call an input channel, input module, channel module, mike channel, mike module, etc. To further confuse matters, name identification of certain alike controls varies from manufacture to manufacture.

The following is a list of sound pickup or sound generating devices you may connect to the input channels of your mixer.

1 – Dynamic Microphone
2 – Electret Condenser Microphone - Requires a battery or "phantom power" from the mixer.
3 – Direct Box - A "direct box" is a device that a musical instrument is plugged into that converts the signal of an instrument's internal microphone or magnetic pickup

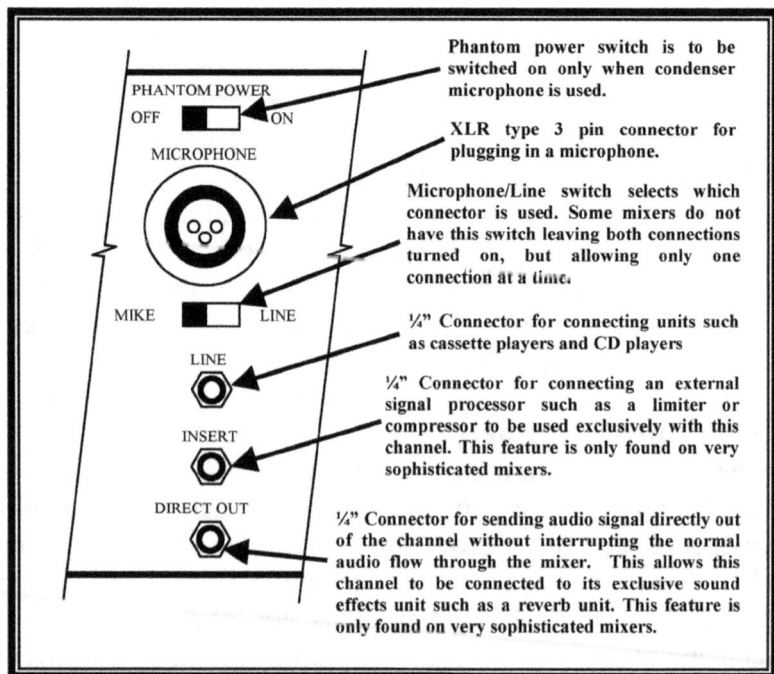

Phantom power switch is to be switched on only when condenser microphone is used.

XLR type 3 pin connector for plugging in a microphone.

Microphone/Line switch selects which connector is used. Some mixers do not have this switch leaving both connections turned on, but allowing only one connection at a time.

¼" Connector for connecting units such as cassette players and CD players

¼" Connector for connecting an external signal processor such as a limiter or compressor to be used exclusively with this channel. This feature is only found on very sophisticated mixers.

¼" Connector for sending audio signal directly out of the channel without interrupting the normal audio flow through the mixer. This allows this channel to be connected to its exclusive sound effects unit such as a reverb unit. This feature is only found on very sophisticated mixers.

PHANTOM POWER
OFF ON
MICROPHONE
MIKE LINE
LINE
INSERT
DIRECT OUT

INPUT CHANNEL CONNECTORS - Rear view of a typical mixer showing connectors and switches to one Input Channel such as found in guitars to a microphone type balanced signal. The "direct box" eliminates the need for a microphone. This box is then plugged into a conventional microphone outlet. This device may also be used with electronic keyboard type instruments and electronic drums.

4 – Wireless Microphone Receiver

5 – CD Player

6 – Cassette Tape Player

7 – The output of an Electronic Reverberation Unit

Only the most expensive and most sophisticated mixers have all the features that have been designed for the INPUT CHANNEL. Manufactures pick and choose what features they deem important in lesser costing mixers. In the following is a list of all the features commonly found on INPUT CHANNELS in the best mixers in the sequence they most often appear electronically. Note that while the features are listed as they appear sequentially, the controls on the INPUT CHANNEL are not. Instead, controls are arranged for convenient operation.

1 – **Microphone Input Connector** - This is an XLR (3 pin) connector for a balanced low impedance microphone.

2 – **Phantom Power Switch** - Turn "on" only when you are powering an electret condenser microphone.

3 – **Line Level Input Connector** - This is usually a ¼" tip/sleeve phone jack to connect line level units such as CD players, tape players, electronic reverberation units, etc.

4 – **Microphone/Line Switch** - This switch selects the input connector to be used. Some mixers do not have this switch leaving both connectors turned on, but allowing only one connection at a time.

5 – **Phase Reversal Switch** - Wiring in a microphone cable consist of two wires, one white and one black, covered by a metallic shield. Sometimes the black and white connections mistakenly get wired backward in a connector on their way to the mixer. When this happens, the microphone's signal will conflict (be out of phase) with another microphone that is wired correctly covering the same sound source. By switching the "phase reversal switch" "on", the cable wiring is corrected electrically.

6 – **Channel On/Off Switch** (sometimes called a Mute Switch) - This switch turns "off" the INPUT CHANNEL when it is not being used.

7 – **Trim Control** (also called Sensitivity Control, Input Attenuator Control, Gain Control and other names) - This control is used to trim the level of the input signals. Often loud sound pickup by a microphone will produce large signals that will over load the INPUT CHANNEL electronics causing sound distortion. Reducing the trim sensitivity eliminates this distortion.

Some mixers use a "–20dB switch" rather than a trim control. When this is switched "on", the INPUT CHANNEL'S input sensitivity is decreased to a set level.

8 – **High Pass Filter Switch** (sometimes called Low Cut Filter) When turned "on", the sound below 100 Hz (G2), called the "roll-off frequency", will be greatly suppressed allowing the higher frequencies to pass unimpeded. Some INPUT CHANNELS have a "Filter Control" to adjust the roll off frequency between about 20Hz (A0) to about 400Hz (G4). The high pass filter is useful to reduce the proximity effect of directional microphones, reduce hum, and reduce stage rumble.

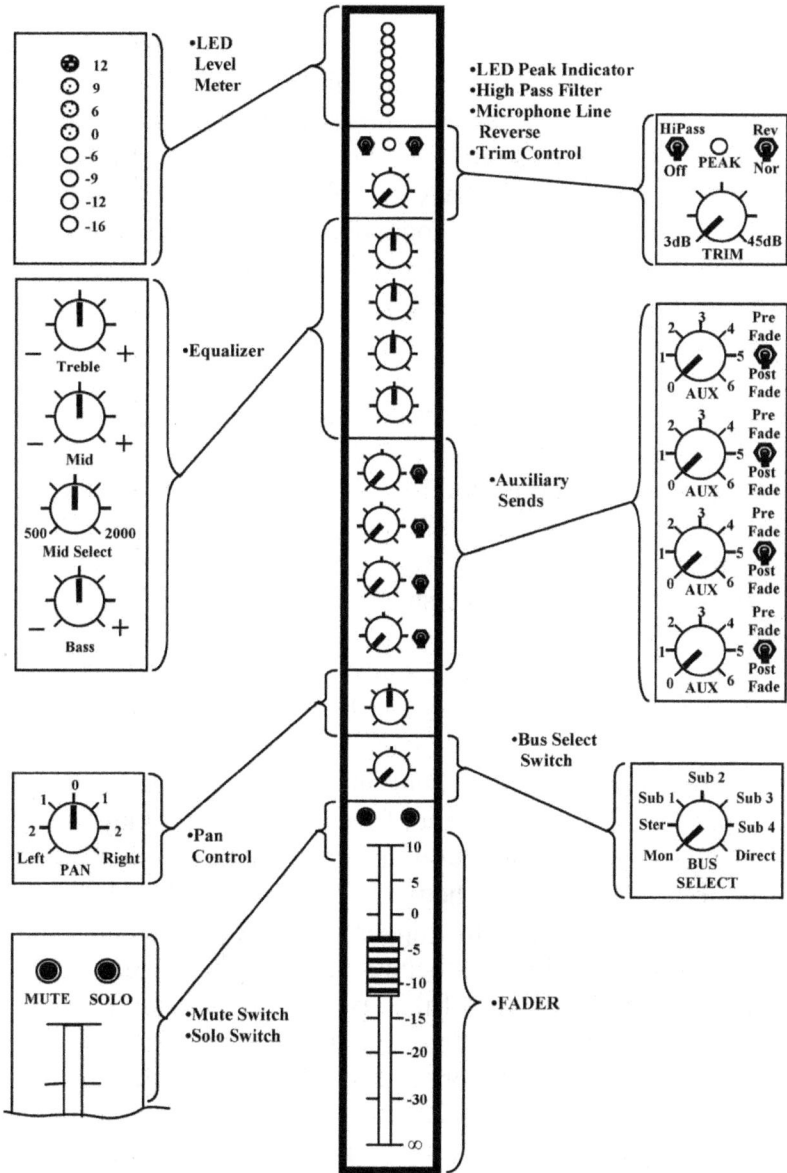

Typical Input Channel Module

9 – **Equalization Controls** – These tone (frequency) adjustment controls usually come in a set of four controls and only adjust the sound signals that flow through that INPUT CHANNEL. The "Bass" and "Treble" controls work similarly to the bass and treble controls on your home stereo unit. In the center position the controls have no effect on the sound. When the bass control is turned clockwise, the bass is boosted and when it is turned counterclockwise, the bass is reduced. The treble control is used the same way to boost and reduce the treble tones.

The tone control for the middle tones work in the same way as the bass and treble controls to boost and reduce the mid range tones, but this control has another associated control that selects the range of middle tones to be adjusted. This control is called a "Sweep" control and you can tune it to a specific part of the sound spectrum that needs adjustment.

These middle tone (mid range) controls are designed to overcome the limitations of the bass control and the treble control. You may want to increase the treble, but find that the treble control adjustment increases too broad a range of high tones. The adjustment of the treble control may give you the clarity you want, but it may also make the sound quality too thin. By returning the treble control to "0" and adjusting the sweep control to the upper tone range, you can find the narrow articulation frequency range without affecting the fullness of the sound. Likewise, you may have too broad a bass boost when you adjust the bass control, and you may find the sound flabby or muddy. Here again you may adjust the sweep control to the lower tone range and find the frequency range that gives you the desired outcome.

Keep in mind, that when you boost the equalization, you increase the loudness a little and sound system feedback might occur from the microphone. It is best to use small amounts of boost when working live. It is frequently more beneficial to reduce the part of the sound spectrum that appears to be overpowering.

Overuse of the bass and treble equalization controls can make things worse. Ordinarily a small amount of treble or bass boost should be sufficient to add brightness or warmth to a sound.

10 -**Insertion Point** - This is usually a ¼″ tip/ring/sleeve phone jack that allows you to connect by cable an external signal processor such as a limiter or compressor to be used exclusively with this INPUT CHANNEL.

11 -**LED Peak Indicator** – This indicator is the best method for ensuring that the INPUT CHANNEL signal is below the clipping distortion level. Typically, if there is occasional flashing on the peak indicator, the signal is not overdriving the channel input. But, if the peak indicator is flashing "on" more than "off", clipping distortion is occurring. This is corrected by reducing the "trim control" sensitivity. Bear in mind that sensitivity to peak signals varies from mixer to mixer and some mixers may show more flashing before clipping distortion occurs than others.

Note that digital mixers are more sensitive to signal clipping and you will want to keep tighter control on the peak signals than you would on the conventional analog mixers.

12 -**Solo (Pre Fade Listen) Button** - When this button is pushed, the sound signal is split off into a headphone circuit where only this INPUT CHANNEL will be heard. The solo button is very useful in troubleshooting problems. It also allows you to make close adjustments with the headphones

before making final fine-tune adjustments by listening to the loudspeakers.

13 -**Fader** - Slide control used to adjust loudness volume on an on going basis is called a fader. It is important to keep your fader around the "0" marked on the fader scale (near mid position) if you can. This is because if your fader position is near the bottom of the scale, a small movement will cause a large change in signal level. Similarly, if you have your fader at the top of its travel, you have no room to boost the signal. Here you may adjust the Trim Control to put the fader in the mid range.

14 -**Auxiliary Controls** (Auxiliary Send Controls) - An INPUT CHANNEL may have only two or as many as eight of these controls. Some auxiliary controls access the input signal that is split off before it gets to the fader and some access the input signal that is split off after it goes through the fader. Then some auxiliary controls have an associated switch that selects the input signal before or after the fader.

When an auxiliary control accesses input signals that are split off before the fader, that is called a "Pre-Fade" auxiliary. Pre-fade auxiliaries are independent and unaffected by the fader adjustments. When an auxiliary control accesses input signals that are split off after the fader, that is called a "Post-Fade" auxiliary. Post-fade auxiliaries "follow" the fader so that when the input level is changed by the fader, the input signal sent to the auxiliary changes proportionally.

Typically, auxiliary controls have two functions. Pre-fade auxiliaries normally feed the monitor or monitor systems to the musicians and post-fade auxiliaries normally feed effects units such as electronic reverberation units.

15 -**LED Bargraph Meter** - Typically, an LED bargraph meter is employed that monitors the output signal of the INPUT CHANNEL.

16 -**Direct Out** - This is usually a ¼" tip/ring/sleeve phone jack for accessing the audio signal directly from the output of the INPUT CHANNEL without interrupting the normal audio flow through the mixer. This allows this channel to be connected to an exclusive sound effects unit such as an electronic reverberation unit by cable. This feature is only found on very sophisticated mixers.

17 -**Pan Control** (Panoramic Control) - Also called a "Balance Control". When the "routing switch" selects the "stereo bus" on the INPUT CHANNEL, this control varies the mono signal level feeds to the "left" and "right" of the stereo bus(es). The pan control sends equal signal levels to the left and right buses when the control is in the center position. Turning the control left will send a higher signal level to the left bus and reduce the right bus signal level proportionally. Likewise when the control is turned right, the right signal level increases and the left decreases.

18 -**Routing Switch(es)** – These switches direct the output signal of the INPUT CHANNEL to the mono bus, the stereo bus, or a sub group bus. When the mono bus is selected, the "pan control" is bypassed.

As the mono bus is selected, the INPUT CHANNEL output signal will be joined with the mono sound signals from other INPUT CHANNELS where the mono bus is selected. Those combined signals will go to the mono "Master-Fader" before being sent to the mixer's "mono output" connector. Likewise, when the stereo bus is selected, the INPUT CHANNEL "Pan Control" signal outputs will be joined with the output signals from other INPUT CHANNELS where "Pan Control" signal outputs are joined when the stereo bus is selected. These combined signals will then go to the stereo "Master-Faders" before being sent to the "stereo mixer output" connectors. The stereo bus is actually two buses that carry the "left" and "right" stereo signals.

There are generally four or more Sub Group Buses in larger mixers and each Sub Group Bus has its own fader. When you select the "sub group 1 bus", the INPUT CHANNEL output signal is joined with the output signals from other INPUT CHANNELS where Sub Group 1 Bus is selected. These combined signals will then go to the "Sub Group 1 Bus Fader" where they are adjusted together. For the convenience of the sound system operator, a Sub Group Bus will allow any number of INPUT CHANNEL signal outputs to be adjusted together on a selected Sub Group Bus Fader. The signals from the Sub Group Bus Fader are then sent to the mixer's Master Faders. Note that on some mixers a single Sub Group Stereo Bus Fader controls both the left and right Channels simultaneously.

Stereo Inputs Channels
Some mixers have STEREO INPUTS CHANNELS that allows you to connect both left and right signals from a stereo source and control them from a single fader. Instruments such as electronic drums and keyboards have stereo outputs that may be directly connected. STEREO INPUTS CHANNELS tend to incorporate fewer features than the mono INPUT CHANNELS because electronic instruments are already equipped with many internal effects and tone options.

Stereo Returns
Most mixers have "stereo return inputs" that are stereo inputs that do not have any features other than a volume level control.

These provide a return signal connection for external effects equipment such as electronic reverberation. You may also use these inputs for connecting and playing a cassette player or a CD player.

Mixer Outputs

The typical mixer will have the following output features.

1. Auxiliary Buses (outputs)– For every auxiliary bus, you will have a signal level control to adjust the combined signals on that bus. The signals on each bus after going through their level control are then sent to individual output connectors on the mixer to feed external effects units and the musician monitoring system power amplifiers.

   Note that pre-fade auxiliary buses (sends) are independent of INPUT CHANNEL fader adjustments and they are used to provide musician monitor loudspeaker sound mixes. You can have as many monitor loudspeaker systems with their individual sound mixes as pre-fade auxiliary buses. It is important to use post-fade auxiliary buses (sends) for effects units such as electronic reverberation. Post-fader auxiliaries "follow" the fader on the INPUT CHANNEL so that when the fader level is changed, the auxiliary send signal on that channel changes proportionally.

2. Mono Bus (output) – All the INPUT CHANNEL mono outputs are collectively adjusted by the mono master fader and sends the signals to the mixer mono output connector for feeding the sound reinforcement power amplifiers that power the reinforcement loudspeakers.

3. Stereo Bus (output) - All the INPUT CHANNEL stereo outputs are collectively adjusted by the stereo master faders and sends the signals to the mixer stereo output connectors.

4. Headphone Output –Headphones are used for listening to individual INPUT CHANNELS and aid in setups and for troubleshooting. A switch is normally provided on each of the INPUT CHANNELS to allow you to monitor their output. Headphones should never be worn to set the sound reinforcement balance.

Mixers for Broadcast or Recording
The sound that you hear in an auditorium is the acoustical sound combined with the boosted sound system sound. If you should record the output of the sound reinforcement mixer you would find the sound mix unbalanced and lacking the sounds of those sources without microphones. This is the reason a second mixer and operator is desired to provide a balanced mix and natural sound for recording or broadcasting.

Here is a typical partial block diagram illustration showing the sound system microphones, a microphone splitter, the sound reinforcement mixer, and the broadcast/recording mixer. As you can see, a microphone splitter (microphone line splitter) allows the output of the stage microphones to feed both mixers. Observe that the choir reinforcement microphones go only to the reinforcement mixer and the broadcast/recording microphones go only to the broadcast/recording mixer. Note also that an audience microphone is connected to the broadcast/recording mixer to pickup congregational singing and the sounds of the auditorium to make a more realistic and natural sounding recording. When placing an audience microphone, avoid putting it in the projection path of loudspeakers or near a loud sound source. Unlike the mixer and operator for the sound reinforcement system, the broadcast/recording mixer and operator should be in an enclosed booth where he or she can see the performance through a window pane or video monitor. The operator needs to be isolated from the sound in the auditorium in order to make a proper mix for broadcast or recording.

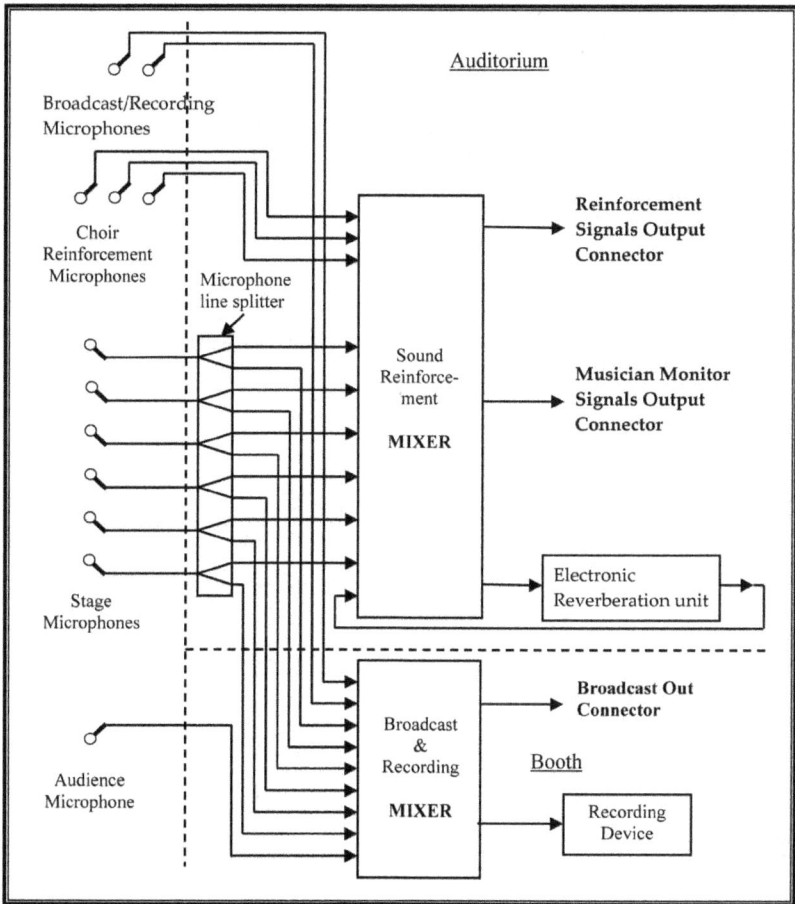

BLOCK DIAGRAM SHOWING REINFORCEMENT AND RECORDING/ BROADCAST MIXERS - Partial block diagram of typical Sound Reinforcement Sound System showing microphones, reinforcement mixer and recording and broadcasting mixer.

Graphic Equalizer

This equalizer is located between the reinforcement mixer and the power amplifier that powers the auditorium loudspeakers. This is called a graphic equalizer because as the front-panel sliders (faders) are adjusted, their positions give an approximate graphic display of the resultant sound spectrum tonal response. Each slider controls a fixed set of narrow band frequencies that may be boosted or cut. Commonly available graphic equalizers have as little as five controls or as many as thirty or more depending or how narrowly the sound spectrum is being divided. The more slider controls you have on your graphic equalizer, the more precise the level of control.

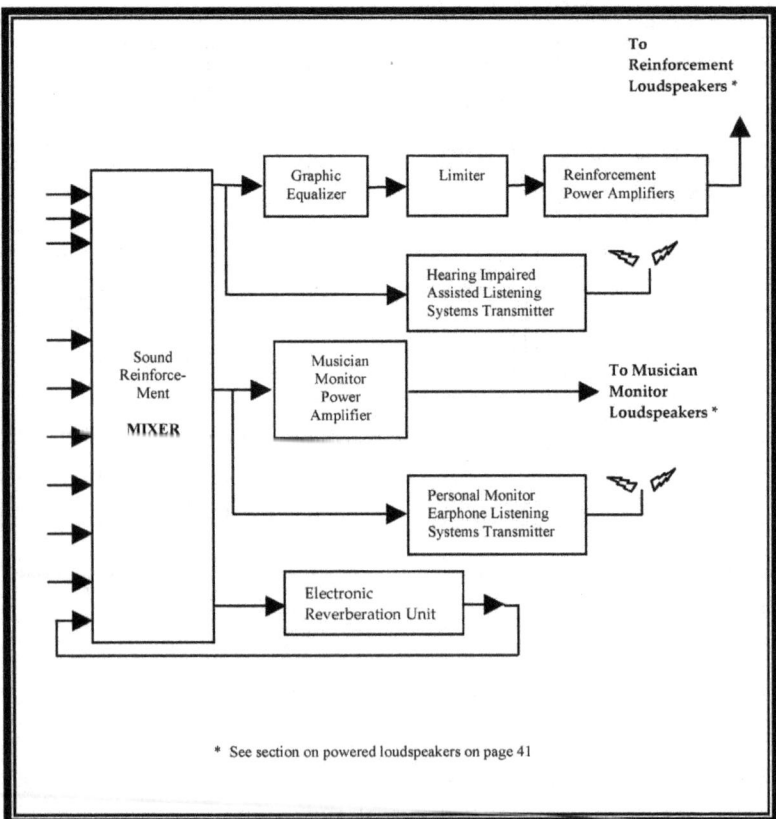

BLOCK DIAGRAM OF SOUND SYSTEM SHOWING THE MIXER OUTPUTS - Partial block diagram of typical Sound Reinforcement Sound System showing the mixer outputs, processing units, hearing impaired transmitter, personal monitor transmitter and power amplification.

The primary use of a graphic equalizer used in sound reinforcement systems is sound feedback control. While they are not the ultimate solution, they can be very useful in this application. The slider adjustments allow you to counter certain room acoustical characteristics by reducing frequencies that tend to feedback.

Compressors

Compressors are typically employed with individual input channels and they may be connected at the input channel's "insert" connector for the purpose of controlling varying sound levels. When someone is singing or playing an instrument, certain notes tend to be louder than others and in a loud environment, quieter passages may get lost unless the operator "rides" the fader to boost the sound during quiet passages. A compressor will allow the input channel fader to be set high enough to hear the low passages while reducing the loud parts so they don't cause distortion.

Threshold, ratio, attack, and release are four controls commonly found on a compressor. The threshold control determines the sound level point the compression function activates and the ratio control sets how much the sound signal is compressed. The speed the compressor reacts to a high level input signal is set by the attack control and the release control sets how fast the compressor releases control after that signal passes. In some compressors the attack and release controls are eliminated and the compressor automatically sets those functions.

Limiters

Limiters are similar to compressors and they are used to prevent signal peaks from exceeding a certain set level in order to prevent overdriving amplifiers, recording tapes and discs. Unlike compressors, limiters are used to remove only occasional peaks and extremely short signal attack and release settings are used so the ear cannot hear the limiter action.

Monitor System for Performers

The purpose for a monitor system is to help the performers on stage. Generally performers cannot hear their voice coming over the sound reinforcement system loudspeakers that cover the audience. Singers and musicians need to hear themselves and each other in order to blend together. Monitor loudspeakers are essential for hearing the music when performances are done with recorded music tracks.

A simple monitor system consists of a power amplifier and one or more distributed loudspeakers. The sound mix to drive the system comes from the pre-fade auxiliary bus output from your reinforcement mixer. Often more than one monitor system is used because performers desire different sound mixes. You can feed as many monitor systems with different sound mixes as you have auxiliary pre-fade outputs on your mixer.

In larger sound systems a monitor mixer is used to feed signals to the monitor speaker systems eliminating the need for pre-fade auxiliary output signal feeds from the reinforcement mixer. The monitor mixer and monitor mixer operator are usually located just off stage to allow easier communication with the performers. See the partial block diagram illustration showing a typical sound reinforcement sound system with a monitor mixer.

There are loudspeaker cabinets especially designed for monitor systems. A "stage wedge" loudspeaker cabinet has two benefits. When it is laid on the floor, the low profile makes it unobtrusive to the audience while pointing the sound toward the performer. The "stand mounted" loudspeaker monitor cabinet is a small unobtrusive cabinet that can be mounted on a microphone stand and placed closer to the performer's ear for intelligibility at a lower sound volume.

It is believed that a more enveloping and well-balanced mix inspires a better performance among musicians and larger "side-fill" loudspeaker monitor cabinets are use with the stage

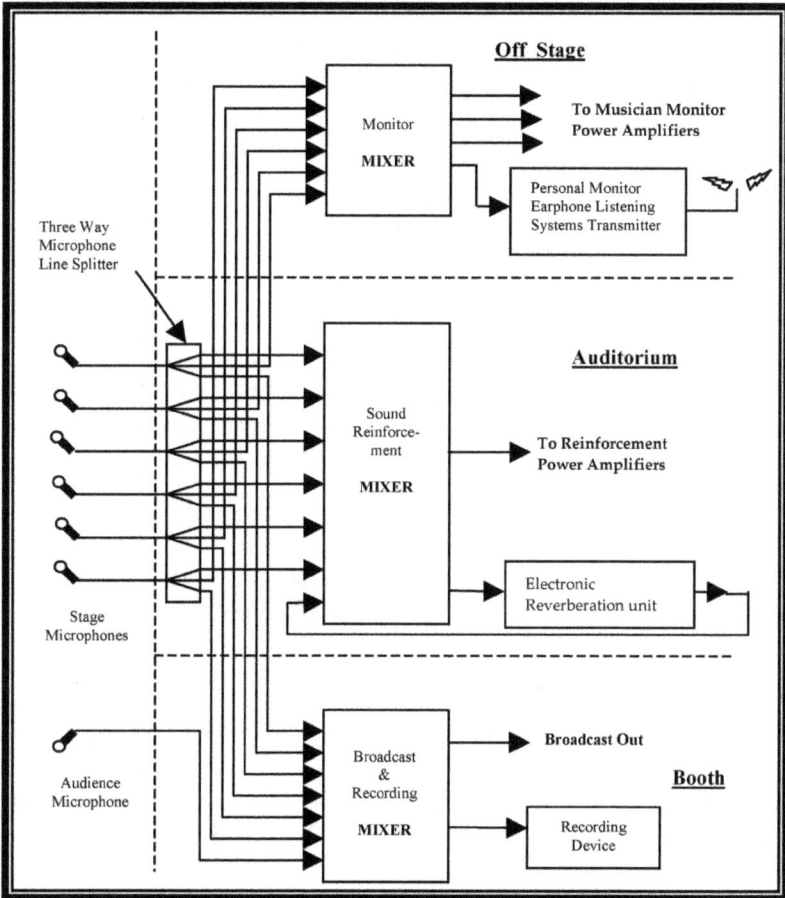

BLOCK DIAGRAM SHOWING A REINFORCEMENT SYSTEM WITH
MONITOR MIXER   Partial block diagram of typical Sound Reinforcement
Sound System showing a monitor mixer.

wedge and stand mounted loudspeaker units to give a fuller sound to the performers. Today, these side-fill loudspeaker cabinets that are placed off to the side of the stage are being used less as technology has made the smaller loudspeakers better.

Personal Monitor Earphone Listening System
Personal monitor earphone listening systems are used to provide sound monitoring directly to the ears of a performer thus eliminating monitor loudspeaker cabinets on stage. Wireless versions are comprised of a transmitter and one or more battery powered bodypack receivers with earphones. As with the conventional loudspeaker monitor system, the sound mix from a pre-fade auxiliary output or monitor mixer provides the signal to the transmitter.

Hearing Impaired Assisted Listening System
Hearing-impaired assisted listening systems provide sound to individuals with hearing impairments. Like the wireless personal monitor earphone listening systems, this system has a transmitter and as many wireless earphone or headphone receivers as necessary. The small receivers are battery operated and the sound heard in these receivers is the same sound signal that is heard from the sound reinforcement loudspeakers.

Electronic Reverberation
When the sound of voices and musical instruments are picked up by closely placed microphones, a "dry" sound may result because little or no surrounding acoustical sound is picked up. To liven the sound, electronic reverberation that creates an acoustical environmental sound is added. The mixer post-fade auxiliary send bus normally feeds the electronic reverberation unit.

Although I have discussed mixers that connect to external reverberation units, many quality mixers contain good electronic reverberation features that gives you the option of using it rather than an external unit.

There are other types of effects units such as delay and echo, but these are mainly used in the recording industry and not recommended for sound reinforcement systems.

Sound System Setup Tips

If you hear a "thump" when you power the sound system, the sound you hear is a possible loudspeaker damaging "switch-on transient". Care should be taken when powering your sound system because your loudspeaker power amplifiers do not have built-in protection circuitry that eliminates this thump. To preserve your loudspeakers, first turn "on" all the other sound system equipment and set the mixer master faders to minimum before you power the loudspeaker power amplifiers. To power down the system, first set the mixer master faders to minimum and turn "off" the power amplifiers before turning "off" the rest of the equipment.

Just as musicians and singers need to rehearse, so does the sound operator with the sound system. The operator should take advantage of the opportunities presented when the choir, ensembles, orchestra, and other groups rehearse. This is a good time to experiment with adjustments and to write down mixer settings so they can be referenced when actual performances are presented.

When more than one person operates the sound system, a methodical approach to setting up the mixer needs to be applied to avoid confusion and surprises during performance time. I suggest that you start by "zeroing" the mixer by moving all settings to a known starting position and adjust as suggested in the following.

1 - Set all mixer input channel faders and auxiliary send controls to minimum settings.

2 - Set all mixer output master faders and auxiliary output controls to their minimum settings.

3 - Adjust all input channel equalizer controls to their flat or neutral position.

4 - Gradually raise the master faders and output controls to the mid position.

5 - Bring up the input channel fader assigned to the lead singer or key musical instrument. Adjust the trim control so the channel fader sets at mid position when a proper sound level is achieved. At this point, should the peak signal indicator still flash "signal clipping", back down the trim control to where the clipping just ceases.

Note that the trim control may have other names such as attenuator, sensitivity, gain, etc.

Establish the maximum working fader level and set it a little below this for a margin of safety before feedback. This will provide a reference level when you bring up the rest of the channels.

6 - If a low frequency background noise is heard, switch "on" the high pass filter except for input channels controlling bass instruments or bass drums (kick drums).

7 - Now adjust all active input channel faders, trims, and high pass filters. Start with vocal microphones and balance them. Now the musical instruments and direct box connections can be balanced with the vocals starting with the drums and

working through the bass and rhythm instruments. As input channel faders are brought up and feedback starts, reduce the master faders until it disappears.

If you have an input channel producing a "hiss" because of a weak source signal, you may be able to reduce it by backing down the trim control and increasing the input channel fader level.

8 - Adjust the equalizers for each input channel. Headphones are useful here to set the equalizer if your mixer has solo (PFL) buttons on the input channels.

9 - Compressors that are used with individual input channels should be adjusted at this point. A compressor is only needed when a person singing or playing an instrument produces certain notes or musical passages too low to be heard properly. A compressor will allows the input channel fader to be set high enough to hear the low passages while reducing the loud parts so they don't overwhelm or cause distortion.

    A - Adjust the threshold control to the point where you want the signal level compression to begin. The quite sounds below the threshold will be amplified normally while the amplification of loud sounds above the threshold point will be reduced. If your compressor has a meter, check with the manual to see how it indicates the compression action.

    B - Adjust the ratio control to set how much the amplification is reduced above the threshold setting. If you set the ratio to 2:1, the amplification above the threshold setting will be ½ of that below. The ratio of 4:1 is ¼ of that below the threshold setting. The higher the ratio, the greater the compression of the above threshold signal.

C – Adjust the attack control to set how fast the compressor reacts to the loud sounds above the threshold settings. The attack is set according to the tonal quality of the sound. The higher the tones, the higher the attack time. If the attack time is set to slow, fast signal peaks will get passed. Many compressors are made with this function automatic and the compressor makes this adjustment for you.

D - Adjust the release control to set how fast the compressor responds to the sudden reduction of loud sound signals above the threshold settings. A very short release time can produce a choppy or jittery sound, and very long release time can result in a squashing sound. All ranges of the release setting may be useful at different times and you should experiment to discover the different sound possibilities. Many compressors are made with this function automatic and the compressor makes this adjustment for you.

10- Mute any microphone input channels not being used.

11- Should you need to adjust your graphic equalizer to "ring out" the reinforcement system feedback, now is the time.

A - Set all slide controls on the graphic equalizer to center zero.

B - Turn the mixer master fader(s) "up" until feedback just begins to "ring", then back the mixer master fader(s) slightly to stop it.

C - Starting from the left, slowly increase the first slider on the graphic equalizer to maximum. If the sound system does not feedback, then return the slider to center zero. Should the sound system feedback when the slider is increased, reduce the slider below zero the same amount it took above zero to start the feedback.

D - Repeat step "C" for the rest of the sliders. Remember that excessive feedback ringing may damage the loudspeakers, so use care when adjusting.

Good automatic feedback reducers may be purchased that will not only achieve the same results as graphic equalizers, but will adjust to changing microphone positions during a performance. They will automatically find and cut the frequencies that cause feedback and they will be effective as long as feedback problems are not increased by microphones being too close to a loudspeaker or poor room acoustics.

12- Adjust the electronic reverberation settings.

13- Adjust your Limiters. Limiters are similar to compressors and compressors will function like limiters when their threshold is set to catch signal peaks and their compression ratio is high. Unlike compressors, limiters are used only to remove occasional peaks and extremely short signal attack and release times are used so the ear cannot hear the limiter action. The limiter's compression ratio is so high the peaks are essentially chopped off.

14- Adjust the stage monitor loudspeaker's mix. Since they can contribute to feedback problems, run these at the lowest loudness the performers can work with.

Glossary

ABSORPTION: the capture of sound into a porous material.

ABSORPTION CAVITIES: tubes or cavities designed to capture excessive bass in a room.

ARRAY: comprised of two or more horn and/or paper cone loudspeakers to provide specific sound coverage in a room.

BOOMY: a splattering or hollow bass sound.

BALANCED LOW IMPEDANCE MICROPHONE: the low impedance microphone electrical circuit design allows for the long distant running of microphone lines (high impedance microphone lines can only be run to about 35 feet without incurring signal losses). Balanced microphone circuits suppress noise and radio interference.

BALANCED LOW IMPEDANCE MICROPHONE MIXER INPUT: in order for the balanced low impedance microphone's electrical circuit to suppress outside electrical interference, it must be connected to a balanced low impedance microphone mixer input.

BAFFLES: boxes to put cone loudspeakers in to control the sound emitted from the rear of the loudspeaker.

CAPACITOR MICROPHONE: another name for condenser type microphones.

COLUMN LOUDSPEAKER: a long slender vertical standing column (baffle) array of cone loudspeakers.

CONDENSER MICROPHONE: a style of microphone that requires an external power supply. It cannot be phantom powered. See electret condenser microphones.

DB (dB): decibel, a unit of measure used to measure sound.

DB-SPL: a unit of measure used to measure sound pressure (hearing) level.

DIFFUSION PANELS: wall panels used to scatter sound.

DIFFRACTION: bending of sound.

DIRECT SOUND: sound not reflected, but heard directly from the source.

DYNAMIC MICROPHONE: the most popular type microphone used in sound systems. It is rugged and does not require a power source like condenser microphones.

EARLY REFLECTED SOUND: sound that reflects rapidly off nearby surfaces that are not distinctly heard apart from the sound heard directly from the source.

ECHO: sound that bounces directly back creating a 1/20 of a second or longer sound behind the direct sound.

ELECTRO/ACOUSTIC SYSTEM: a sound system that has loudspeakers projecting sound into an acoustical environment.

ELECTRET CONDENSER MICROPHONE: a style of microphone that requires a battery or an external power source. It can be phantom powered by a mixer that has that feature.

FEEDBACK: the returning of loudspeaker sound to a live microphone causing a squealing sound when the amplification in the sound system is too high.

FREQUENCY: number of vibrations or cycles per second.

HERTZ (Hz): a cycle per second.

LED: "Light-Emitting Diode".

LOUDSPEAKER ARRAY: a selection of two or more horn and/or paper cone loudspeakers grouped together.

MID RANGE OF TONES: range of tones or frequencies from about C3 (130Hz.) To C7 (2,093Hz.).

PINK NOISE: sounds like a waterfall and contains all the tones (frequencies) uniformly produced that man can hear.

Glossary

POWERED LOUDSPEAKER: are generally combination horn and cone loudspeaker cabinets that contain the power amplification necessary to drive the loudspeakers. Since the power amplifier is in the cabinet (baffle), an AC power outlet must be close by.

REFRACTION: when sound vibrates a wall, a window or some other similar structure causing the sound to be radiated to the other side, then the sound is said to be refracted.

RESONANCE: ringing sounds in areas of a room that are stimulated by certain tones.

REVERBERATION: is created when sound reflects many times off multiple surfaces such as hard walls, a hard floor, a hard ceiling, and hard objects within a room and is the diminishing sound that remains after the direct sound is heard.

ROOM: any structure containing the acoustical environment considered. The room may be an auditorium, a sanctuary, a meeting room, etc.

SOUND PRESSURE LEVEL METER: measures sound pressure rather than sound power or intensity.

SOUND PRESSURE: is a way to measure sound that relates to the way we hear.

STANDING WAVES: vibrating air made by a tone projected into a room is called a wave train. This wave train striking a wall will produce a reflected wave train. Standing waves are created when two wave trains moving in opposite directions interfere with each other.

WHITE NOISE: similar to pink noise except the loudness increases slightly as the frequency increases.

XLR CONNECTORS: typical of the three pin connectors used on low impedance microphones.

Appendix 1

Sound and Acoustics Handbook

Appendix 1

Some Thoughts on Listening and Mixing

In the 1970s I met two gentlemen who taught me how to listen to music. James R. (Jim) Combs had been a radio disc jockey for several years and Bill Browning was a former Rockabilly recording artist and songwriter who had hit records in the 1950s. We worked together in Midway Recording Studio. Jim was the audio recording engineer and Bill a producer. I maintained all the electronics and multi-channel recording equipment, and helped in the studio during recording sessions. Here I learned a great deal about recording techniques and acoustics. By observing and listening to these men, I obtained sufficient knowledge to become a recording engineer.

Listening to recordings and discussing what comprised good reproduced sound never ended. Never being content to stop improving what we produced kept us experimenting and listening.

The first thing I learned was to listen to hear all the instruments individually and distinguish the tonal characteristics of each. Often, a producer may say to a recording engineer during a recording session, "The bass guitar is too boomy", or, "The brass is too sharp", or maybe, "Bring up the rhythm guitar, I don't like the balance in the rhythm section". At this point the engineer may adjust the EQ or sound level on a microphone channel. The recording may be stopped to readjust, or even change a microphone. At times, the producer or recording engineer might ask a musician or singer to make an adjustment in performing position or instrument setup.

A producers main concern is with creating the most pleasing overall sound. He or she does not care how the controls are adjusted on the recording console.

Do not mix so that all the sound level meters read the same sound level intensity. You must listen and mix to get a proper balance. The level meters are there to show you when you are reaching distortion levels.

Generally, rhythm instruments are reviewed and blended first. Then major accompanying instruments and lead voices are added. Afterwards the rest of the voices are mixed and following that, the rest of the instruments.

When mixing sound for a sound reinforcement system, you don't have the advantages that a recording studio provides. But, this process of listening and initially mixing in stages can serve you well. That is why working with rehearsal performances is essential where you can chart your mixer setup for each production. A chart showing levels, EQ settings, reverb levels, bus assignments, etc., for each production can be very helpful. Gum labels on microphone channels to identify the microphone connected is a must if many microphones are being used.

Remember, in order to be a good sound mixer, you must be able to individually hear each sound produced. Then you must be able to blend those to a proper mix. While maintaining a pleasing sound you must be able to mentally scan each individual sound frequently while you are listening. Amplified sound with a bit of a bright edge usually sounds best. Do not allow sound to become muddy sounding because of to bassy a mix. Also, note that in a reinforcement system you are mixing to blend with the room's acoustical sound.

Do not become frustrated if your skills do not develop rapidly. Becoming a good technician takes time and diligence. Know your sound equipment. Listen, mentally take apart, and analyze what you hear on the radio and all the media sources where you hear performances. Never stop learning.

Appendix 2

Soldering

Faulty microphone and instrument cables with wires that break through much handling cause the most failures in a sound system. Learning to repair your cables will save you time and save you the frustration that comes from relying on someone else.

Almost all cord problems occur at the cable entry to the connector or at a solder joint inside the connector. Cords are flexed the most at the connector and wires break at that point. Also, cables that are not tightly anchored at be connector entry will allow wires to be stressed at the solder joints.

Solder Tool Kit
A solder tool kit should include the following articles.

1. 35 – 100 watt pencil type soldering iron with holder
2. De-soldering pump
3. 60/40 (60% Tin-40% Lead) flux cored solder
4. Soldering iron tip cleaning sponge
5. "Helping Hand" type soldering aid
6. Wire insulation strippers
7. Small diagonal wire cutters
8. Small long nose pliers
9. Regular pliers
10. Flat blade screwdriver
11. Phillips screwdriver
12. Set of inexpensive jeweler type flat blade screwdrivers
13. Set of inexpensive jeweler type Phillips screwdrivers
14. Small sharp knife or single edge razor blades
15. Safety Goggles

Solder

Solder used in electronics is 60% Tin and 40% Lead (60/40) and the core of the solder wire contains a paste cleaning material called flux. When soldering, the flux melts before the solder to clean wire and terminal to be joined.

Be sure you buy solder made to be use on electronic circuits. Acid core solder will damage electronic circuits. Solder may be purchased without flux, but to use it, you will need a container of flux paste and you will have to apply it manually.

Soldering Irons

Inexpensive soldering irons are usually adequate for occasional use and may last for many years if used infrequently. They will last as long as the soldering tip last, then they are discarded. Solder stations (soldering irons with temperature control) for light use sell for less than $50.00. These allow for tip replacement.

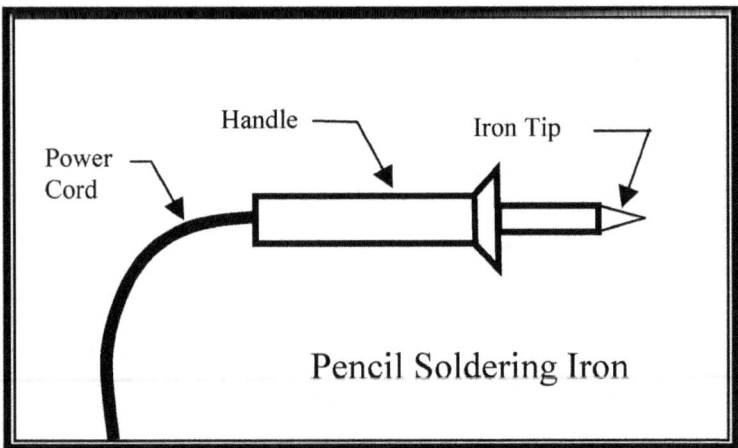

Handle · Iron Tip · Power Cord

Pencil Soldering Iron

De-Soldering Pump

Before repairs to a connector are made, the old solder must be removed. A small inexpensive hand held spring-loaded de-soldering pump works well. This device is cocked and applied to a heated terminal. Pushing a trigger activates it and it sucks the old hot solder from the terminal. De-soldering pumps are also commonly referred to as "solder suckers".

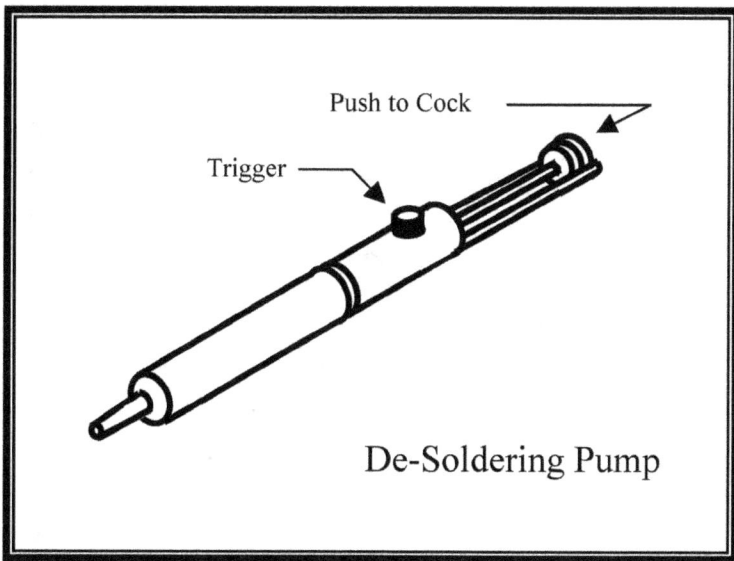

Push to Cock

Trigger

De-Soldering Pump

Soldering Holding Aid

Holding a connector, a wire, and an iron while soldering is difficult. I find a "helping hand" soldering aid or small vise to hold the connector is beneficial to get the job done easily. These can be bought with a heavy base for stability, a suction base, or a tabletop clamp.

Wire Insulation Strippers

Wire Insulation Strippers are hand-held devices used to safely strip the insulation off the end of a piece of wire. The most common strippers resemble a pair of long nose pliers with a series of various sized sharpened half-holes cut into the jaws that match up with each other when the jaws are closed. Here you select the hole size that corresponds to the solid or stranded wire size to be stripped. To strip, you close the hole around the insulation and pull it from the wire. Be careful to select the right hole and not pinch or scrape the wire when stripping.

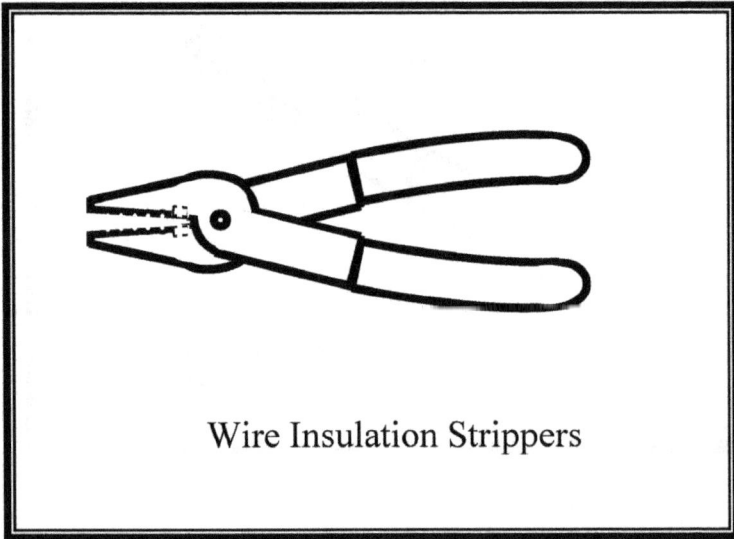

Wire Insulation Strippers

Another kind of wire stripper has an adjustable diamond-shaped hole in the jaws, and they can be set to close at any size. Note also that a wire cutter is included on both these type strippers sufficient not only to cut wire, but also microphone and instrument cables.

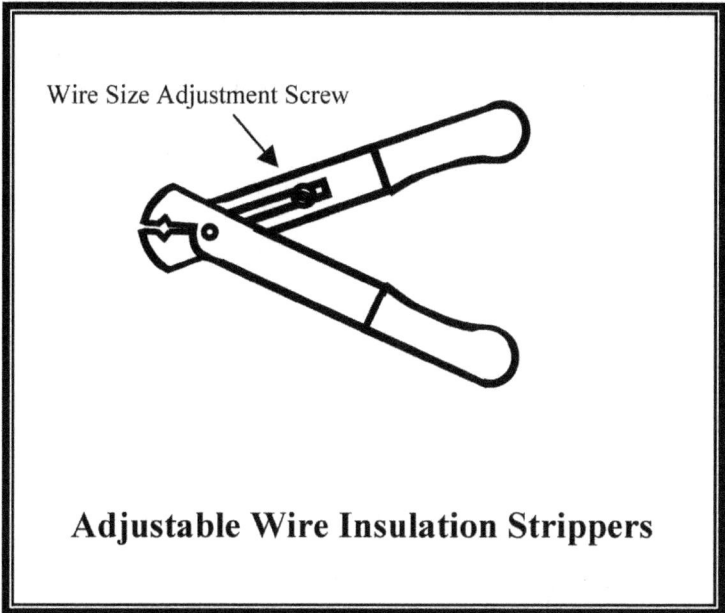

Adjustable Wire Insulation Strippers

Automatic wire strippers have a mechanical linkage powered by a spring which can grab, cut, and pull the insulation automatically as the handles are pressed closed. These are a bit costly and are more than you need for light duty work.

Small Diagonal Wire Cutters
These small wire cutters are used to cut and trim small electronic wire. Cutting any wire other than soft copper or tinned copper wire will ruin them. They may break if you attempt cutting cable or heavy wire.

Small Long Nose Pliers
Long nose pliers are used to twist electronic wire around solder terminals. Heavy clamping and grasping items too large will damage the pliers.

Jeweler Type Screwdrivers

XLR microphone connectors and phone plug connectors found on instrument cords are made by many different manufactures. Some manufactures assemble their connectors with tiny screws making jeweler type screwdrivers the ideal tools. These may be purchased from an electronics parts store or from a place that sells small hand tools. Note that some Switchcraft type XLR connectors have a tiny screw with "left hand" threads.

Sharp Knife or Single Edge Razor Blades

A knife or razor is used to carefully remove the insulating jacket from the cable you are repairing. When removing do not damage the shield. The shield may consist of fine wire braid, fine wire wrap, or mylar film with a bare wire.

Safety Goggles

Occasionally solder flux may pop from a joint you are soldering like hot grease from a frying pan. Wire fragments will sometimes be propelled when wire is trimmed. Always wear safety goggles or glasses.

Tinning Iron Tips

Before using an iron, the tip needs to be cleaned and "tinned". A hot tip is best cleaned on a wet sponge. Sponge holders with sponges are sold for this purpose and sometimes are supplied with an iron or a solder station. The iron tip is designed with a metallic coating that will accept solder. When a hot iron is cleaned, apply solder to the tip and allow the tip to become shiny (tinned). The soldering iron is ready to use.

Soldering

In order for a satisfactory solder joint to be formed, the wire and terminal on the connector must be clean and the right amount of heat applied. Normally the cleaning action of the flux in the core of the solder is sufficient for a proper solder joint. The soldered surfaces must be totally free of corrosion before melted solder will adhere properly.

Two types of solder terminals are almost always found in sound system cable connectors. These are solder-cup and lug.

Before soldering the wire to a connector with solder-cup type terminal, tin both the wire and the terminal lightly with hot solder. Both the wire and terminal should be shiny. If the wire is stranded, tightly twist the wires together before tinning so that there will be no frayed wires. Allow about one eighth inch wire between the wire insulation and the terminal.

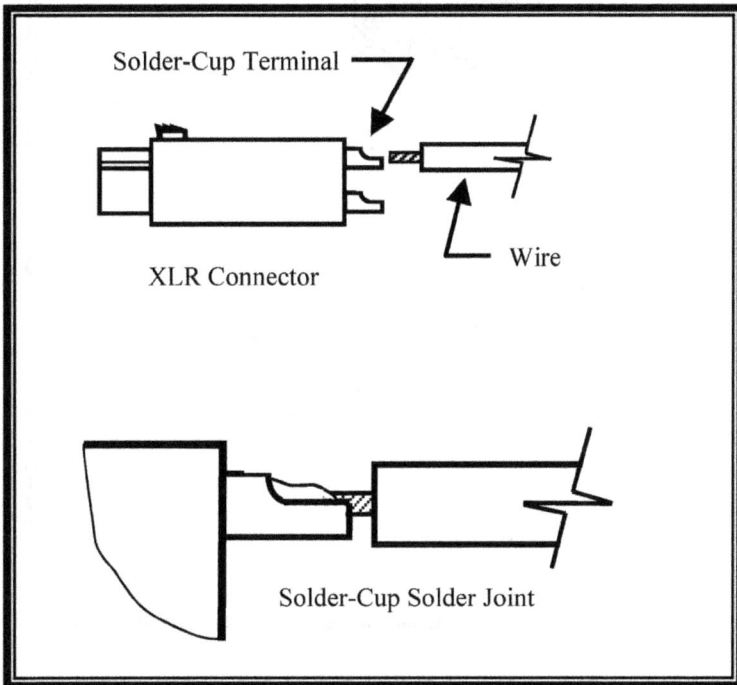

Solder-Cup Terminal

XLR Connector

Wire

Solder-Cup Solder Joint

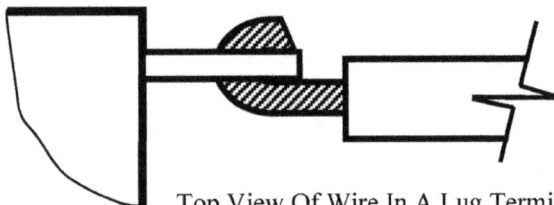

Top View Of Wire In A Lug Terminal

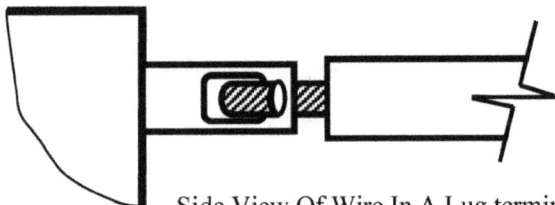

Side View Of Wire In A Lug terminal

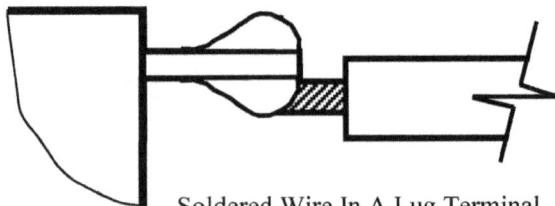

Soldered Wire In A Lug Terminal

When working with connectors that have solder-cup type terminals, slip the tinned wire into the tinned cup terminal then apply the tinned hot iron tip to both the wire and terminal simultaneously. Immediately apply the cored solder to the wire and terminal very close to the iron tip allowing the solder to flow smoothly. Sufficient solder should fill the cup while showing the contour of the wire.

If you are soldering a wire to a lug terminal, put the tinned wire through the lug hole and bend the wire into a ¾ to 1 turn loop. Crimp the wire tightly with your long nose pliers to the lug. Apply the tinned hot iron tip to both the wire and terminal simultaneously. Immediately apply the cored solder to the wire and terminal very close to the iron tip allowing the solder to flow smoothly. Solder should fill the area where the wire and lug come in contact while showing the contour of the wire. Note that lug type terminals are not normally tinned before soldering. The hole in the lug does not need to be filled.

The soldering action for both of these type terminals should take about three seconds. Do not use excessive solder that leaves domes, peaks or an overflow.

If the iron is not sufficiently hot or not applied long enough to the wire and terminal, a "cold" solder joint will result leaving a dull looking glob of solder. A properly soldered joint will be shiny and smooth.

Avoid any movement of soldered components until the solder hardens or a fissured joint will result. Fissured and cold solder joints can cause electrical intermittence and noise.

It is very important to remember to tin the soldering iron tip at the end of each session, otherwise the tip alloy coating will deteriorate causing the tip to rust. If this happens, the iron may not work very well or at all.

Appendix 3

Charting Your Mixer Settings

Just as musicians and singers need to rehearse, so does the sound operator with the sound system. The operator should take advantage of the opportunities presented when the choir, ensembles, orchestra, and other groups rehearse. This is a good time to experiment with adjustments and to write down mixer settings so they can be referenced when actual performances are presented.

When more than one person operates the sound system, charting your settings may be necessary in order for you to be able to maintain control quality. I find a clipboard to hold notes and charts works well.

You can create a chart by sketching the controls of your mixer on a sheet of paper. Make your original drawing the master sheet and make copies of it. It may not be practical to draw all the controls on one sheet of paper. If you can clearly draw 12 microphone channels, use two copies if you need to chart 24 channels. Make your chart easy to mark and easy to read. Do not put in any unnecessary information. Make it as simple as possible.

Should a backup operator need to stand in for you, charts made for a performance would be invaluable.

Depending on the complexity of your mixer setup, you may also want to make a chart for your mixer out put controls.

In the following is an examples of how to make and mark charts. Note, reference the mixer microphone channel on page 81. The chart is based on this microphone channel.

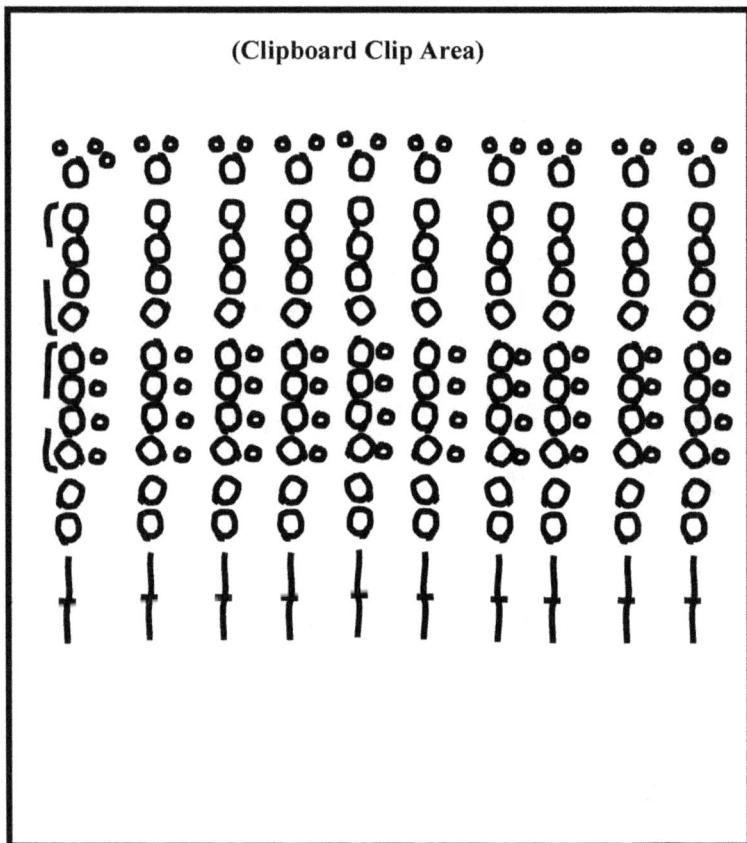

(Clipboard Clip Area)

Typical Mixer Control Sketch to Record Settings

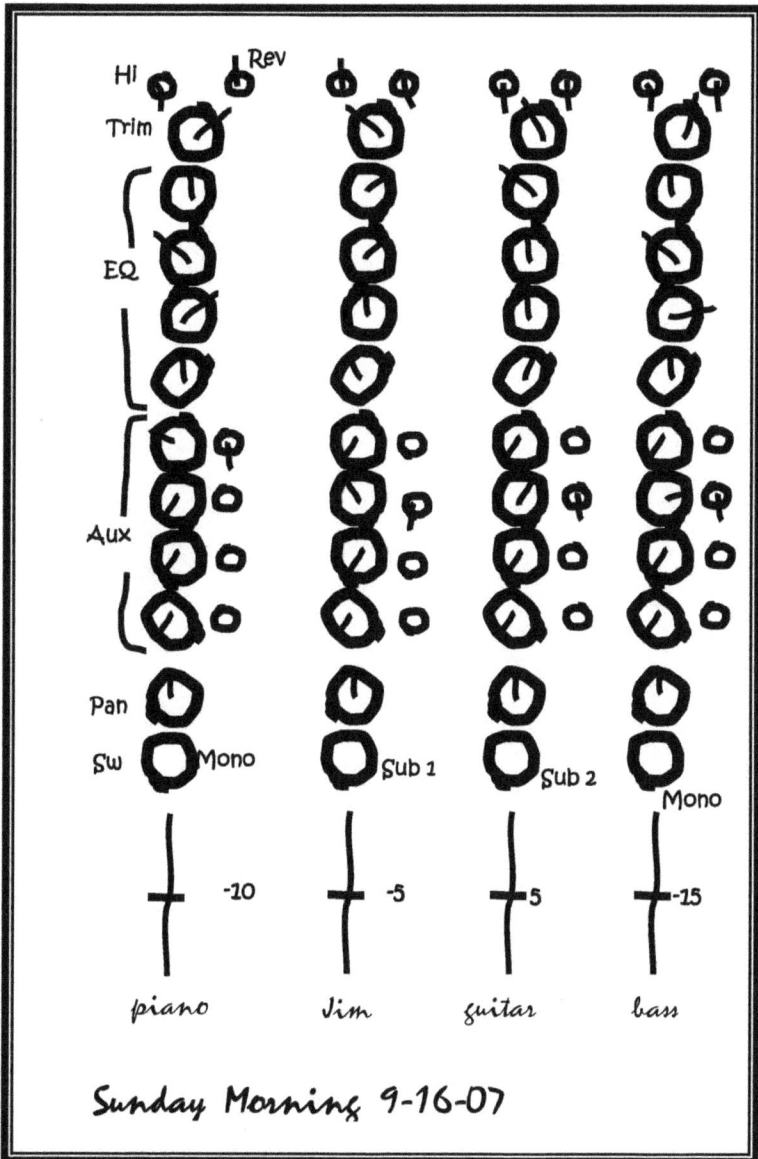

Slash Marks And Notes On Chart To Indicate Settings

www.ingramcontent.com/pod-product-compliance
Lightning Source LLC
Chambersburg PA
CBHW061736020426
42331CB00006B/1261